TELECOMMUNICATIONS AND DEVELOPMENT IN AFRICA

Telecommunications and Development in Africa

Edited by

B.A. Kiplagat and M.C.M. Werner

Telecommunications Foundation of Africa

1994

IOS Press

Amsterdam • Oxford • Washington DC • Tokyo

ISBN 90 5199 169 X

Publisher:
IOS Press
Van Diemenstraat 94
1013 CN Amsterdam
Netherlands

Sole distributor in the UK and Ireland:
IOS Press/Lavis Marketing
73 Lime Walk
Headington
Oxford OX3 7AD
England

Distributor in the USA and Canada:
IOS Press, Inc.
P.O. Box 10558
Burke, VA 2209-0558
U.S.A.

Distributor in Japan:
Kaigai Publications, Ltd.
21 Kanda Tsukasa-Cho 2-Chome
Chiyoda-Ku
Tokyo 101
Japan

PRINTED IN THE NETHERLANDS

FOREWORD

A CHALLENGE TO POLITICIANS, FINANCIERS AND ENGINEERS

The Telecommunications Foundation of Africa (TFA) was created in 1992, out of a conviction that insufficient telecommunications in Africa are an impediment to economic growth and that more African and international resources could be mobilised to strengthen this sector. Political will to achieve progress is as important as more creativeness of financiers, regulators and engineers.

TFA made this volume for readers inside and outside Africa, and more importantly perhaps, for readers both inside and outside the telecommunications industry. We are offering ideas and propositions on financing, regulation, use of advanced technology and rural telecommunications. Our primary point of reference are users, both locals and multinational organisations, since we treat telecommunications as a key stimulant to economic growth.

The telecommunications sector in Africa is more resourceful than many outside and inside this continent are aware of. There are many astute telecommunications managers and excellent engineers in Africa. There is also a solid installed base of state-of-the-art technology in many countries.

However, what is holding the continent back are political restraints on the role of the private sector, lack of autonomy of telecommunications operating companies and insufficient regional cooperation. Furthermore, there is inadequate information available on the African telecommunications sector. Neighbouring operators inside Africa lack insight in each other's progress, which prevents countries from identifying issues of mutual interest. Even within the boundaries of a single country, information is often insufficient or inaccessible, as one contribution illustrates.

Once such impediments are removed, financial resources will become readily available, since as we all know, telecommunications is good business. As is evident in other continents, telecommunications will then also produce more foreign exchange than ever before. Sector reform in other continents is showing a consistent growth, from which the existing monopoly operators also benefit due to increasing traffic on their networks. Without private sector participation in the telecommunications domain, those monopolies usually do not have the human and capital resources to develop niche markets. In addition, evidence suggests that investment in telecommunications has a stronger effect on economic growth than any other infrastructural investment.

Requirements of international business users are as important in Africa as anywhere in the world, as is basic telecommunications service. However, prevailing economic, financial and regulatory orthodoxy are still failing to offer satisfactory recipes, neither for better services for business users, nor for creating universal access to telecommunication services. Telecom planners and economists are often incapable of balancing costs and benefits of low density telecommunications in urban and rural areas, especially in view

V

of the rapidly dropping line costs for rural solutions. For example, the real economic benefits of a rural payphone can be many times higher than an urban residential connection. Likewise, national and regional backbone networks could accommodate more advanced services, once they are upgraded with the help of outside capital. Much more research is required in these areas and TFA is prepared to be actively involved.

This volume puts forward a number of important proposals and we would like to highlight the following:

(1)   Telecommunications companies could be given autonomous powers to raise capital internationally, without seeking prior government approval.

(2)   Private capital can be mobilised after legal reform of the telecommunication sector. Such reform will enhance both the leverage of existing telecommunications companies and create new possibilities for supplementary service providers. These measures will result a in bigger penetration and a wider portfolio of telecommunications services, the latter being of particular importance to international companies.

(3)   Telecommunications should be considered as part of development programmes in other sectors, notably agriculture, education and health. This will be instrumental in solving the deadlock in advancing rural telecommunications, which is one of the biggest challenges this sector faces in Africa.

(4)   Advances in rural and low density telecommunications technology necessitate review of the economics of providing services to remote and disadvantaged areas. In many cases, telecommunications regulations may well leave room for self-financed independent regional and local operators.

This publication views Africa in a global perspective, in economic, regulatory and technological terms. Experiences in regulatory development in Europe are important, especially in respect of the multigovernmental framework in which this development is taking place and also given that the initiative for change has been associated with macroeconomic drives rather than with established telecommunications operators. Arguments are offered for ensuring that Africa keeps pace with global technology: whereas the rest of the world is gearing towards multimedia communications and the associated productivity gains. Technological and architectural choices in Africa's networks should not preclude upgrading in due time.

A number of individual countries are presenting their experiences in corporate restructuring and achievements in network expansion. So far, country cases where sector reform is facilitating competition and permitting new entrants in the telecommunications services market are still few in number and limited in scope. Some countries have issued *ad hoc* government licenses for cellular radio telephone services. However, experience shows that incidental licensing prior to putting a comprehensive regulatory framework in place which addresses a multi-operator environment could lead to legal and accounting problems later. Independent National Telecommunications Regulatory Authorities are in place in only very few countries. The creation of such bodies deserves the highest priority at governmental level, since this is often a key to attract the best players for introducing new and better services in a country, in cooperation and alongside existing operators.

We wish to thank all authors most sincerely for their contributions and also express our gratitude to the Government of the Netherlands for co-financing the production of this publication.

Bethuel A. Kiplagat
Marcel C. Werner

editors

All authors are contributing on a personal basis. The views they express are not necessarily the views of the organisations they belong to nor do they necessarily reflect TFA positions.

TABLE OF CONTENTS

PART I

ECONOMICS, FINANCE AND REGULATION

Telecommunications and Development in Africa
B.A. Kiplagat and M.C.M. Werner, Eds.
IOS Press

CHAPTER I

CONSTRAINTS IN FINANCING TELECOMMUNICATIONS IN AFRICA'S LEAST DEVELOPED COUNTRIES

Kordjé Bedoumra
Chief Engineer Telecommunications
African Development Bank
Côte d'Ivoire

SUMMARY

Although it is universally acknowledged that telelcommunications is a profitable sector, most African countries are unable to mobilize the financial resources needed to develop it. The situation is very critical for the least developed countries whose weak economies considerably hinder their efforts to gain access to capital at market rates. We therefore urgently need to reflect on ways and means of developing telecommunications in these countries. Since telecommunication is a capital-intensive sector, no stone should be left unturned if we wish to attain an acceptable growth level. We can start by establishing an appropriate legal framework, then proceed to attract the private sector, and finally by integrating telecommunication infrastructure components into social and/or priority projects.

INTRODUCTION

In light of the present context in which Africa is beset by chronic macro-economic and financial imbalances essentially marked by:

❖ adverse terms of trade engendering a drastic fall in prices of exported raw materials;
❖ an ever-increasing debt burden;
❖ a shrinking of the flow of capital.

There seems to be very little hope of raising sufficient capital to increase the meagre investments made in the telecommunications sector in Africa.

Today communications is at the heart of all economic activities and constitutes a prerequisite to the development and efficiency of all other sectors of the economy. The technological evolution is opening up unsuspected horizons for new services capable of contributing towards economic growth.

This article merely sets out the problem of mobilizing resources for financing the telecommunications sector, particularly in the least developed countries. It begins by elaborating on the profitability of the telecommunications sector and then reviews the constraints to the further mobilization of resources for this sector. The last part of the article bears on possible solutions to the problem.

PROFITABILITY OF TELECOMMUNICATIONS IN AFRICA

According to certain statistics (1), the turnover per main line is two and a half times higher in some African countries than in certain developed countries. Since the rates and traffic per line are relatively higher in relation to other regions of the world, telecommunications in Africa is particularly attractive financially. In spite of the comparatively heavier investment unit costs, the operating fixed assets always show a profitability of more than 10%, with a few exceptions due to poor management. The internal financial profitability rate of telecommunications projects often exceeds 15%.

Although the development of telecommunications is closely linked to the economy of the country, telecommunications can be managed profitably and efficiently even in low per-capita income countries. This is amply testified by such countries as Cape Verde, Burkina Faso and Senegal. For instance the Senegalese telecommunications company SONATEL was able to make an advance repayment of some of its debts, ensure self-financing up to 64% and reduce useage tariffs by 31.6% for firms and 21.8% for residential subscribers (2).

Generally speaking, the interest rates applied by bilateral and multilateral institutions on loans intended for telecommunications do not exceed 10% and can therefore be borne by most telecommunications projects. However, the funds raised for financing telecommunications are still inadequate.

THE CURRENT SITUATION REGARDING FINANCE

It can be seen from the data on the operations of the two main development financing institutions in Africa, i.e. the African Development Bank and the World Bank, that their involvement in telecommunications has been relatively modest in relation to the needs. Over a three-year period, between 1990 and 1992, the World Bank invested approximately US$ 250 million in this sector, i.e. about US$ 83.3 million a year (3). During the same period, the African Development Bank Group granted loans amounting to about US$ 324.5 million for telecommunication projects, i.e. an average of US$ 108.2 million per year (4). The third major multilateral source which provides finance for telecommunications in Africa is the European Development Fund. The other financial backers are mostly bilateral, comprising in particular German, Canadian, French, Italian, Swedish and Japanese cooperation programmes.

Investment in telecommunications in Africa is estimated to be less than US$ 2 billion a year, whereas to attain a density of 3.11 main lines per 100 inhabitants in the year 2000 (as against the present density of 1.4), Africa would require an annual investment of US$ 4.493 billion between 1992 and 2000 (5), i.e. more than double the current investment level.

Multilateral development finance institutions have several financing windows, operating mainly on normal terms and soft terms. For instance, the African Development Bank has a relatively large quantity of funds provided on normal terms with an interest rate currently hovering around 7%, and soft loans from the African Development Fund (interest-free loans over 50 years) for the least developed countries. One might be tempted to think therefore that with its high profitability rate, the telecommunications sector could have easy access to non-concessionary loans for its development. This however is not possible for many African countries.

CONSTRAINT TO ACQUISITION OF FINANCE

The multilateral finance institutions only lend to Government or with the latter's guarantee. This implies that the whole debt situation of the country is taken into consideration, regardless of the viability of telecommunications projects. The structural adjustment programmes and the debt-relief measures have failed to stem indebtedness, which continues to be an unbearable burden for many countries. Furthermore, 33 low income African countries, boasting a per capita annual income of less than US$ 510 (6), find it virtually impossible to obtain funds on non-concessionary terms from development finance institutions. They can only seek soft loans whose volume is restricted by the fact that they are made up of voluntary contributions provided by developed countries on the basis of their capacity and policies. As a result of the recent devaluation of the CFA franc, the number of countries concerned will probably increase from 33 to 37. The Governments of these countries are devoting their meagre resources first and foremost to the survival of their people (food, water, health, etc.) and the development of human resources.

In order to develop their telecommunication networks, the least developed countries mainly rely for the moment on either bilateral aid or soft loans from multilateral institutions. This aid alone is inadequate for a rapid development of telecommunications as evidenced by the virtual non-execution of many telecommunication development master plans.

Thus if the present state of affairs continues, many African countries will not be able to develop a selfsustained sector such as the telecommunications sector, capable of generating income (taxes, ect.) and indispensable to the economic growth of the country.

In view of the difficulties African Governments are encountering in their efforts to borrow or guarantee further loans, we must seek other approaches for stepping up the mobilization of capital necessary for financing telecommunicayions in this region.

PREREQUISITE TO THE RESTRUCTURING OF TELECOMMUNICATIONS

The prerequisite to any mobilization of resources is the sound management of the company. This can only be achieved in a favourable macro-economic environment. There have been many documents and conferences on the restructuring of telecommunications. What we shall do here is to merely recall a number of commonly acknowledged recommendations:

I   The role of the State is to ensure that public services are provided, whether the network operations are public or private, and that the rules of competition are respected where necessary. The following are the three major requirements to be fulfilled:
a) meet demand within reasonable time and over the whole territory;
b) sell the services at the least cost;
c) provide quality service.
The State should have appropriate mechanisms for ensuring that these goals are met and the rules of the game are respected.

II   Public telecommunications companies will have to be granted real financial and administrative autonomy necessary for running a private enterprose.

WAYS OF FINANCING TELECOMMUNICATIONS

FREEDOM TO BORROW WITHOUT STATE GUARANTEE

The telecommunication company's financial autonomy should include the freedom to borrow without State intervention. The State guarantee would only come in where the donors demand such guarantee. However, the freedom to borrow would not imply operation of the telecommunications network(s) without any State control. Indeed, the borrowing would be carried out in the framework of objectives laid down jointly with the State and within debt limits acceptable to the company. Moreover such operations would be submitted to the telecommunication company's Board of Directors on which the State would also be represented. It would therefore be useful for the legal personality granted to the telecommunications company to provide for the right to borrow, with the constraints that this might entail for the quality of its management.

Access to capital markets should be achieved freely even for public telecommunication companies. In this regard we could point to the experience of France Télécom with respect to the
financing of its needs (7). Although this is a public company, France Télécom borrows under its own name. It finances its short-term loans through credit lines with banks as well as enjoys the possibilities offered by the national telecommunications fund. The long-term loans are financed mainly on the financial market by borrowing on international franc markets. As for its operations abroad, France Télécom resorts to international markets in Euro-currencies.

While it is true that many African countries lack capital markets, it is also just as true that the universalization of the economy is opening new horizons and the creation of regional development poles should offer better prospects. There are examples of cases where shares have been issued on stock exchanges in the West African region: Abidjan, Accra and Lagos. These are possibilities which should be explored further.

PRIVATE CAPITAL: SEIZE OPPORTUNITIES

In view of the consirable needs in investment and the low volume of capital available, it is increasingly essential to have recourse to private capital. To do this we must establish a legal and institutional framework that will enable the private sector to take part and thereby contribute to a real long-term development of telecommunications.

In the context of the areas laid down by the law, the regulating body should assist and encourage the private sector and ensure that the interests of all parties are defended equitably. This body should be able to effectively defend the public's interests, making sure that investments are made at the least cost, the techniques used are the most suitable, the rates are financially and economically reasonable, the quality of service is acceptable and the users are treated fairly. It is therefore important that this body be endowed with the requisite professional personnel in the technical, financial, economic and legal areas. This personnel should be adequately remunerated and be granted terms of employment which guarantee it a minimum of independence.

Private sector participation should not be restricted solely to the issue of opening up to the private sector the capital of the public telecommunications operator or only to the introduction of competition for new services. The limited extension of the basic network calls for greater imagination in efforts to attract private capital. It has been noted in some African cities moreover, that some private or professional subscribers are willing to take part in the financing of the investments needed for acquiring telephone and some new services. Such an opportunity is not seized systematically and furthermore, when the case does happen, the subscribers often do not receive the necessary compensation.

There are several issues such as the extension of the basic network in certain geographical areas or the use and maintenance of part of a network, which could be entrusted to the private sector, in the framework of agreements fair to all parties and satisfactory to users. It is therefore a question of getting the private sector involved at all levels without necessarily privatizing beforehand the public operator of telecommunications network.

It has been observed from the experience of some African countries, that when a public telecommunications enterprise is privatized, the opening up of the majority capital to a private foreign partner does not per force lead to a mobilization of resources. Indeed sometimes the industrialized countries, other parties than the private partner concerned, stop their contribution to the country's telecommunications sector. Furthermore, the partner in question does not provide substantial resources and does not raise additional private capital. There is therefore a need to exercice much more caution when entertaining the idea of offering majority shares of the telecommunications company to a private external partner as a way of attracting more private capital.

Success in mobilizing private capital depends heavily on the country's economic and legal context: exchange and facilities and possibility of transferring capital abroad, bank regulations and access to credit, promotion of the private sector, etc. This is what leads multilateral financial backers to include the reform of the telecommunications sector in the context of macro-economic adjustment and restructuring of the whole parastatal sector. Telecommunications cannot develop singlehandedly in an environment which is unfavourable to business in general.

FINANCING FROM INTERNATIONAL DEVELOPMENT INSTITUTIONS

Although finance from multilateral development sources will continue to come in, given the limited capital available on soft terms, contributions will always be small in volume. This constraint does not in any way diminish the decisive role which these institutions should play in the restructuring of the sector. The resources could be used first and foremost to support countries in their efforts to introduce and strengthen appropriate reforms. The accumulated experience of these restitutions coupled with their mission, arms them more effectively to help countries find appropriate solutions for developing their telecommunications.

INTEGRATION OF TELECOMMUNICATIONS IN SOCIAL PROJECTS

All development partners recognize the need to extend telephone services to the rural areas where the majority of the population lives. Most Governments require telecommunications companies to include rural telecommunications in their investment programmes. However this can only be done to a limited extent and in a slow manner, because the unit costs in these areas are much higher for lighter traffic and yet the company must maintain a healthy financial balance.

It is mainly the economic factor rather than the financial aspect which justifies telecommunications in rural areas. Most of the rural population in the least developed countries clearly lacks the means to subscribe to and pay for a telephone, but in order for development projects set up to assist them to succeed, there is need to establish telecommunication networks. The needs in communication include both audio and visual aspects.

The growing number of possibilities offered by modern communication tools and their ever decreasing cost makes it possible to conceive of them and integrate them into agricultural, health or educational development projects and programmes in which their costs would be marginal.

It is therefore high time that the design of the above-mentioned type of projects start to include the setting-up of telecommunications facilities, which are essential for those projects to be effective. This would be in the form of a component of an agricultural, educational or health project for the rural population concerned. Therefore, soft loans can also be earmarked for telecommunications. The role of the national telecommunications operating company will be to ensure that technical standards are respected and that sound interconnection with the public network is established. This operating company could then be given the responsibility of maintaining and operating such remote and small networks. It is mainly up to telecommunication companies to make management in the various sectors concerned aware and convince them of the technical possibilities offered to them by present day telecommunications.

CONCLUSION

The African countries' indebtedness has considerably hampered their capacity to borrow and this has adversely affected such highly profitable sectors as telecommunications. It is therefore imperative that other ways of coming out of this impasse be found.

The administrative and financial autonomy to be granted to public operators of telecommunication networks should include the power to borrow without necessarily having to obtain the usual State guarantee.

In addition to recourse to bilateral and multilateral financing sources, there should be a major, well-organized and monitored campaign, aimed at acquiring assistance from the private sector, in order to accelarate the development of telecommunications.

Those (local/regional) telecommunications infrastructures which are not financially profitable, but which are economically indispensable to the success of programmes in other vital sectors such as agriculture, education or health can be integrated into these vital sector's projects.

OoOoOoO

REFERENCES:

(1)   Attentes et vues des investisseurs dans le contexte spécifique de l'Afrique au Sud du Sahara - François Verges - I.C.E.A. Exposé au Colloque sur "Telecommunications et prospérité économique: une stratégie pour l'Afrique" Abidjan, Côte d'Ivoire, 21-25 février 1994.

(2)   Politique et stratégie nationale du secteur des télécommunications au Sénégal-Cheikh Tidiane Mbaye, Directeur Général SONATEL. Exposé au Colloque sur "Telecommunications et prospérité économique: une stratégie pour l'Afrique" Abidjan, Côte d'Ivoire, 21-25 février 1994.

(3)   The Bank's Experience in the Telecommunications Sector, November 2, 1993, World Bank, Washington D.C.

(4)   Banque africaine de développement, Fonds africain de développement, Rapports annuels, 1990, 1991, 1992.

(5)   Indicateurs des Télécommunications africaines, BDT-UIT, 1993.

(6)   Fonds Africain de Développement-Politique du FAD VI-Octobre 1991.

(7)   France Télécom - Les télécommunications - Dirigé par François du Castel - X,A Descours- Berger-Levrault International.

Telecommunications and Development in Africa
B.A. Kiplagat and M.C.M. Werner, Eds.
IOS Press

CHAPTER II

THE PARADOXES OF AFRICAN TELECOMMUNICATIONS

Michael Minges, Tim Kelly
International Telecommunication Unit (ITU)
Telecommunication Development Bureau (BDT)
Switzerland

By most measurements, telecommunications in Africa lags behind every region in the world. It has only two per cent of the world's main telephone lines despite having 12 per cent of its population. It had the lowest annual growth in teledensity (main telephone lines per 100 inhabitants) of any developing country region over the last ten years, partly due to rapid population growth. Some 35 of the world's 49 least telecommunications-developed countries are African.

Figure 1 : Regional trends

Average annual growth rate in main lines and reledensity, 1983-1992 and distribution of main lines by region, 1992

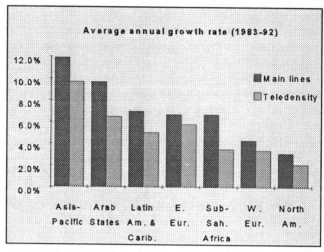

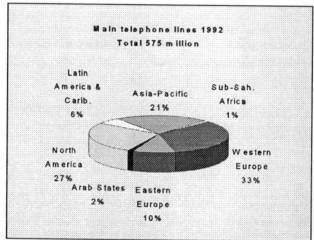

Source : ITU World Telecommunication Development Report, 1994.

Factors external to the telecommunication sector explain part of the problem. Foreign debt is seriously hampering the capability of countries to import telecommunication equipment. Population growth is high so that network development never seems to catch up with demand. Civil disturbances in some countries damage the already fragile telecommunications infrastructure and creates an environment which is hostile to investment confidence.

At the same time, the telecommunication sector within the region is characterised by a number of paradoxes which suggest that not all of the problems are external to the sector. One paradox is that the level of international telephone traffic per subscriber in Sub-Saharan Africa (213 minutes per year) is the highest of any region in the world, but the level of traffic per inhabitant (less than one minute) is the lowest in the world. This suggests that there is considerable demand but insufficient supply for telecommunications within the region.

The second paradox is that the average level of pre-tax profitability of public telecommunication operators in Africa is among the highest of any region in the world (almost 34 per cent of revenue in the case of North Africa). On the other hand, the level of revenue per inhabitant is less than any other region, (just over US$ 5 per year in Sub-Saharan Africa). Thus telecommunications is indeed profitable in the region, perhaps because access to telecommunication services is limited to the richest fraction of the population and to the foreign-owned, export-oriented sectors of the economy.

A third paradox is that, even though the average level of income in Sub-Saharan Africa is the lowest of any region in the world, nevertheless the cost of installing a telephone line, a somewhat labour-intensive activity, is the highest of any region in the world. The average cost of each new line, calculated by dividing the value of investment in the public telecommunication sector by the number of new lines installed during the year, is more than US$ 5'500 per line. This compares with a developing country average of just under US$ 1'500 per year. If the Sub-Saharan Africa figure was closer to the world average, the region would be adding four times as many lines a year. This would translate into almost 600'000 extra lines per year that could potentially be created through more efficient investment procurement based on best-practice from elsewhere in the world.

Figure 2 : Busy lines but few callers (Paradox 1)
Minutes of outgoing international telephone traffic in selected African countries 1992, per inhabitant and per subscriber

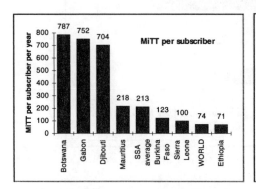 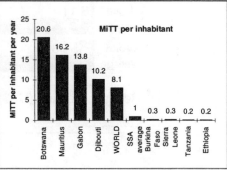

Note : MiTT = Minutes of Telecommunication Traffic. Outgoing international traffic is used in this figure SSA = Sub-Saharan Africa
Source : ITU/BDT Telecommunication Indicator Database.

**Figure 3 : Profitable operators but low revenues per inhabitant (Paradox 2)
Pre-tax profitability of public telecommunication operators and revenues per
inhabitant in Africa and the World, 1992.**

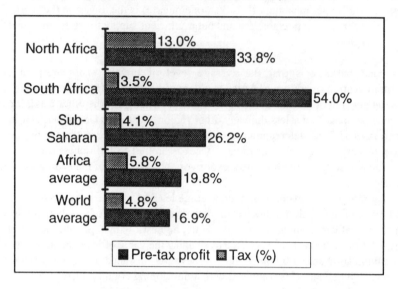

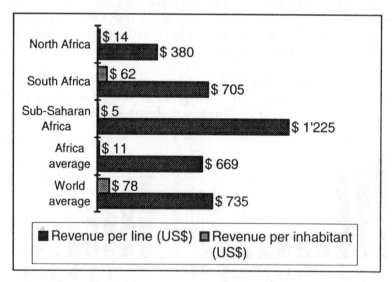

Note : Pre-tax profit is defined as operational surplus after debt payments but before tax
and before exceptional items.

Source : ITU/BDT Telecommunications Indicator Database.

Figure 4 : Cheap labour but expensive lines (Paradox 3)
Level of Gross Domestic Product per capita and cost of adding new lines, in
Africa and the World, 1992

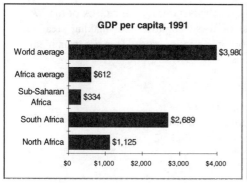

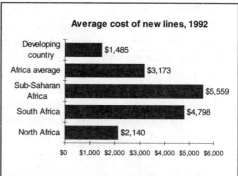

Note : The cost of new lines is calculated by dividing capital investment in 1992 by the
number of new lines installed in that year.
Source : ITU/BDT Telecommunication Indicator Database.

What these paradoxes suggest is that problems associated with telecommunication
development in Africa are not strictly caused by external factors or a lack of money.
Rather, many problems could be considerably alleviated by structural reform within the
telecommunication sector itself and which need not wait for changes within the rest of
the economy. Some countries have already achieved this, notably Botswana, Egypt and
the Gambia. These countries have consistently outperformed the regional average
throughout the last decade. This shows that the problems of the telecommunication
sector can be tackled providing there is sufficient political will. It also illustrates the value
of separation between telecommunication and postal functions and between the operator
and the state as a first step towards reform. This has been carried out from an early date
in each of these countries.

NETWORK DEVELOPMENT

There is a high level of unsatisfied demand for telecommunication service in Africa. On
average, there is only one telephone line for every 235 people in Sub-Saharan Africa; this
is around 120 times less than in a developed country. There are almost four million
people "officially" waiting for a telephone line which is equivalent to around a third of
the existing subscriber base. The number of these official "waiters" is growing over 7 per
cent a year. There are undoubtedly millions of others who want a telephone line but have
not bothered to register. Newer services such as cellular, radio-paging and packet switch

data networks have yet to be implemented in a majority of countries in the region. The low level of existing subscribers combined with increasing unsatisfied demand, a fast growing population, and a lack of new services suggests that the situation will get worse unless rapid telecommunication network development can be achieved.

A number of countries in the region fared well over the last decade. In fact six of the 17 fastest growing networks are African. Most of these countries are making steady progress and moving up in terms of relative world penetration rates. Two are examined in more detail to see what lessons can be learned.

Table 1 : High achievers
African economics with per annum growth rate in main lines
of over 15 per cent per year, 1983-92

	Annual			Teledensity Average		Teledensity rank	
Cape Verde	22.7%	0.61	3.14	19.9%	133	109	24
Burundi	20.6%	0.05	0.23	17.1%	182	166	16
Gambia	19.9%	0.33	1.33	16.9%	149	129	20
Botswana	16.1%	0.95	2.62	12.0%	126	117	9
Egypt	15.7%	1.32	3.94	12.9%	117	102	15
Chad	15.4%	0.02	0.07	12.6%	183	182	1

Note :   Not including economies with a population under 100,000.
Source : ITU/BDT Telecommunication Indicator Database.

BOTSWANA

Botswana is listed among the Least Developed Countries recognised by the United Nations Development Programme (UNDP), but its telecommunications infrastructure is one of the most modern and extensive in Africa. Some 36'500 subscriber lines are in operation and all are connected to digital exchanges. That gives a teledensity of 2.6 lines per 100 population which is the third highest in sub-Saharan Africa. The network has grown at a rate of 20 per cent per year since 1987. The Botswana Telecommunications Corporation (BTC) is investing in a fibre optic backbone network due for completion in 1994. This will link the main cities and will supplement the existing microwave inter-exchange network.

Clearly Botswana is not representative of developing countries as a whole. Its economy is based on diamonds and other natural resources, and it has a growing income from tourism. It also has a small population, 1.4 million, spread over a very wide area. Nevertheless, there are certain steps which Botswana has taken since independence which could be emulated by other developing countries. These include:

❏ Separation of the PTO from the State. Even though BTC is government-owned, it was formally separated from the state in 1980 and is managed by Cable and Wireless.

❏ BTC maintains a high quality of service with some 80 per cent of all faults repaired by the next working day.

❏ BTC has continued to invest at a high rate. Investment as a percentage of revenue has been consistently above 40 per cent throughout the last decade and in 1992 investment actually exceeded revenue, in preparation for the introduction of the new packet-switching and radio-paging services.

❏ While concentrating initially on providing services to the business community, BTC is now widening its reach. In 1992, for the first time, more than 50 per cent of all lines served residential subscribers and the number of payphones installed grew by 10 per cent. BTC has publicly announced a policy of providing at least one pay phone in every village of the country with more than 500 population.

EGYPT

In the early 1980s the Egyptian telecommunication network was underdeveloped. Teledensity was around one, there were 300 pay phones serving a population of over 40 million and only 15 cities were connected to the long distance network. In 1982, the Arab Republic of Egypt National Telecommunication Organisation (ARENTO), the state-owned telecommunication operator, was granted considerable autonomy to carry out its Five year plan (1982-87) for improving telecommunications. Ten years later the results speak for themselves: teledensity had almost quadrupled to four, payphones had increased by a factor of 10 to 3'700 and the number of cities with access to the long distance network had grown to over 300.

The autonomy granted ARENTO allowed it to implement a number of policies to improve performance. These included introduction of an incentive scheme for staff, a strategic planning function, and decentralisation by delegation of authority to the appropriate department. ARENTO has carefully monitored implementation and follow-up of its plans by tracking quantitative indicators on a monthly basis. Deviations from targets are acted upon quickly.

ARENTO also adopted a number of steps to generate revenue for investment in the network. These included raising installation fees, introducing time limits for local calls, charging business users higher subscription fees and introducing a priority fee for subscribers requesting immediate access to the network. As a result it has self-financed an increasing portion of investment with less reliance on government loans. ARENTO reinvested around 60% of revenues from 1982-93. Sources of hard currency for financing equipment imports included low interest bi-lateral loans guaranteed by the Egyptian government and supplier credits. Foreign currency was also earned from the high volume of incoming international telephone and telex traffic due in part to the large number of Egyptians abroad.

FINANCING NETWORK INVESTMENT

The difficulties of financing network investment is a perennial problem for the public telecommunication operators of Africa. There are a number of reasons for this:

◻ **THE LACK OF FOREIGN EXCHANGE TO PURCHASE TELECOMMUNICATION EQUIPMENT.** Given the lack of local equipment manufacture, at least outside South Africa and some of the countries of North Africa, most equipment has to be imported and paid for in hard currency. Import taxes add further expense. The recent 50 per cent devaluation of the CFA Franc in the Francophone countries of West Africa will further disadvantage network investment in these countries by making imports relatively more expensive;

◻ **EARNINGS RETAINED BY THE STATE.** Even though most of the operators of the region make an operational profit, sometimes quite substantial, and even though few of them pay tax, nevertheless the government retains a high share of earnings because it actually owns the operator in most African countries. This problem is particularly severe for incoming settlement payments from foreign operators which are often retained by the Ministry of Finance and are rarely earmarked for network development.

◻ **HIGH COSTS FOR NEW LINES.** Line costs in Africa are generally higher then in other developing countries. This can be partially explained by the low volume of purchases which raises unit costs, by lack of local manufacture and because line costs are usually higher in the early stages of network development. However, high line costs are also partially due to inefficient procurement practices and the lack of competitive tendering. A recent report analysing the telecommunications sector in a West African country states: *"foreign loans were tied to foreign credits, hence the public telecommunication operator had little say in the choice or cost of equipment for development projects"*.

◻ **INEFFICIENT TARIFF STRUCTURES.** Tariff structures in many African countries are inefficient as a way of raising revenues. International calls provide a much higher share of overall revenue in African than in other regions but potential revenue is lost because of high call failure rates, shortage of outgoing international circuits and because the high international tariffs.

Despite these funding problems, the available evidence does seem to suggest that the fundamental difficulties are organisational, not financial. The level of television ownership, for instance, is more than twice as high in Africa as the level of telephone ownership. This is because television ownership is basically a private investment decision, made by individuals, with no waiting list, whereas telephone ownership is a public investment decision, made by state-owned corporations, for which the waiting time is, on average, more than five years.

The secret, therefore, may be to make telecommunication more like a private investment decision and to leverage, as much as possible, private sources of funding. The introduction of mobile communications provides a good example of this. Once a basic national infrastructure is established, the decision of whether or not to buy a mobilephone rests with the user, not with the supplier. Furthermore, mobile communications has generally proved to be a source of revenue to the state (through the sale of licences and through a tax on call charges and mobilephone sales) rather than an added investment burden on the State.

MULTILATERAL DEVELOPMENT ASSISTANCE

It will be some time before the majority of African countries can fully implement reforms to attract sufficient private investment in the telecommunication sector. The development of capital markets in Africa able to finance telecommunication development on the scale needed will also take some time. The multi-lateral development banks could assist in the interim by providing financing and linking it to structural reform. The support of the multi-lateral agencies -- who provide hard currency loans -- is especially important in Africa where declining exchange rates and high foreign debt levels leave little resources for financing telecommunication equipment imports. Furthermore, they could enforce open tendering which should lead to lower equipment costs.

The African Development Bank (ADB) is the region's multilateral development bank. Based in Abidjan, Côte d'Ivoire, it has 51 regional members and 25 non-regional members. It has loaned over US$ 25 billion since 1967 for economic development projects in the region. Its stated policy on infrastructure, including, telecommunications is: *"Regional member countries have, since independence, attached an important priority to investments in infrastructure; and the Bank Group's loan portfolio indicates that infrastructural development projects have accounted for a significant portion of its lending in areas such as transport, water supply, energy and telecommunications. ... in view of inadequate communications ... in many parts of Africa, the Bank Group continued to support projects aimed at addressing these weaknesses."*

The record on ADB support for telecommunications is mixed. Over the ten year period from 1983-92, some US$ 260 million was provided for 12 telecommunications projects. This amounted to 1.2 per cent of total lending which is less than the telecommunication sector share of African GDP. ADB telecommunication lending declined from 1983-86, there was no lending from 1987-89 and lending has see-sawed since then.

The ADB record on telecommunications is weak when compared to other multi-lateral lenders. The World Bank lent almost 3 times more to the region for telecommunications while the European Investment Bank lent almost as much for telecommunications to the

region as the ADB. The Asian Development Bank, -- similar to the African Development
Bank in terms of size and total lending -- spends twice as much on telecommunications
as the ADB. The new European Bank for Reconstruction and Development, which has
total loan portfolio one tenth the size of the ADB's, has nonetheless lent almost twice as
much for telecommunications projects in two years than the ADB did in ten.

All of the multi-lateral development agencies might do more for Africa. Though the
African region received the second largest share of multilateral financing for
telecommunications from 1983-92 (around 25 per cent amounting to US$ 1.3 billion),
Eastern Europe was a close third and most of that amount came in the last two years.
Furthermore, over half the telecommunication lending to Africa went to just three
countries: Algeria, Morocco and Nigeria.

Figure 5 : A bigger slice of the pie ?
African Development Bank (ADB) telecommunication lending as a percentage
of total lending, 1983-92

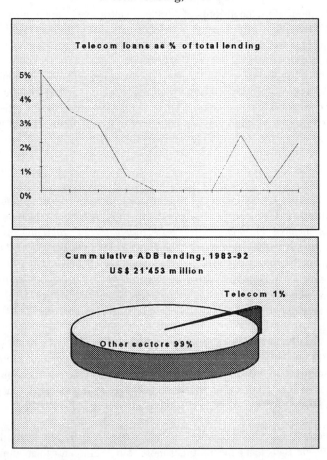

Source : ITU adapted from ADB reports.

Table 2 : Multilateral help (1)
African Development Bank lending for telecommunications, 1983-92

Year	Country	US$	Description
1983	Angola	28.9	Improvement of the quality of international telephone and telex services through the extension and modernisation of the Cacuaco earth satellite station.
1983	Malawi	8.8	Expansion of telex and telegraph and extending these services to the whole country.
1983	Zimbabwe	6.7	Assist the Post and Telecommunications Corporation in improving telecommunication service including provision of underground cables, overhead wires and insulators, switching equipment, subscriber telephone apparatus, and other equipment.
1984	Ethiopia	2.5	Extending telecommunications service to semi-rural areas; expanding services in small urban centres; reducing the telephone service waiters and strengthening maintenance.
1984	Mozambi-que	26.4	Rehabilitation of the country's most important transmission system and providing adequate telephone services in and around the port city of Beira which also serves the landlocked countries: Malawi, Zambia and Zimbabwe.
1985	Zambia	30.8	Rehabilitation and expansion of the existing network. The major components of the project comprise provision and installation of local exchanges and expansion of cable distribution.
1986	Senegal	9.1	Rehabilitation of existing telecommunications network; provision of facilities for maintaining the network; and improving the quality of service.
1990	Angola	31.6	Improve quality of service offered to users, open up the provinces and contribute to the modernisation, extension and development of the telecommunications network.
1990	Zaire	42.3	Set up efficient management in the Postal Services and Telecommunications sector as well as secure for the sector a minimum of reliable means of communication to pursue the rehabilitation of the installations in the city of Kinshasa.
1991	Malawi	10.5	The project seeks to modernise the telecommunications network of the capital Lilongwe and the Municipality of Zomba including the peripheral areas of these two centres. To do this, the project provides for the supply and installation of digital exchanges.
1992	Ethiopia	36.6	Expansion of telecommunications services in order to meet present and future demand by increasing the number of telephone subscribers and upgrading rural services.
1992	Tanzania	23.6	Improve the quality of services of the network, as well as increase the telephone penetration rate.
	TOTAL	257.7	

Source : ITU/BDT Telecommunication Indicator Database.

Table 3 : Multilateral help (2)
European Investment Bank lending for telecommunications in Africa, 1983-92

Year	Country	US$	Description
1983	Zimbabwe	13.4	Improvements to trunk and international telephone and telex
1985	Congo	4.6	Improvement of international telephone connections
1985	Kenya	16.8	Extension and modernisation of local and trunk telephone networks
1986	Côte	9.8	Rehabilitation and reinforcement of national and international
1987	Senegal	13.9	Expansion and modernisation of national telecommunications
1988	Togo	0.1	Preparatory study on implementation of projected digital radio
1989	Benin	7.7	Expansion of telephone system serving Cotonou and of international
1989	Rwanda	8.8	Extension and improvement of domestic and international
1989	Togo	10.3	Expansion and improvement of telephone networks in Lomé and
989	Zimbabwe	19.8	Rehabilitation and expansion of telecommunications network
1992	Ethiopia	7.8	Rehabilitation of northern section of telecommunication network
1992	Morocco	103.7	Upgrading telecommunications capacity with Europe by laying
1992	Senegal	16.9	Renewal and extension of telecom facilities in northern Senegal
	TOTAL	**233.6**	

Source : ITU/BDT Telecommunication Project Database

Table 4 : Multilateral help (3)
World Bank lending for telecommunications in Africa, 1983-92

Year	Country	US$	Description
1983	Uganda	22	Uganda I
1984	Algeria	128	Algeria I
1985	Ethiopia	40	Ethiopia VI
1985	Kenya	32.6	Kenya III
1986	Côte	24.5	Côte d'Ivoire
1986	Senegal	22	Senegal II
1987	Burundi	4.8	Burundi II
1987	Morocco	125	Morocco I
1987	Tanzania	23	Tanzania II
1989	Benin	16	Benin I
1989	Ghana	19	Ghana II
1989	Togo	16	Togo I
1989	Uganda	52.3	Uganda II
1990	Nigeria	225	Nigeria I
1991	Rwanda	12.8	Rwanda II
	TOTAL	763	

Source : ITU/BDT Telecommunication Project Database

Telecommunications and Development/TFA

MOBILE COMMUNICATIONS

The potential for cellular communications remains largely untapped in Africa. Until 1992, cellular was only available in the North African countries (except Libya), Mauritius and South Africa. However, the tide appears to be turning. In 1992 four more countries commenced operation of cellular systems and others are about to begin operations or are planning.

One way of quickly introducing cellular is through joint ventures with established foreign operators. This is being done to some extent in Africa. For example Millicom International Cellular, a Luxembourg-registered company with worldwide cellular operations, has established joint ventures in Ghana, Mauritius and Tanzania. Vodaphone and Cable & Wireless, two British companies, were recently awarded licenses to operate digital cellular (GSM) networks in South Africa. A US company -- Digital Telecomms -- established a cellular joint venture with the Nigerian Telecommunications Company (NITEL) in 1992.

Many developing countries have adopted cellular as a quick solution for by-passing long waiting lists and low fixed-line availability. This could be especially relevant in Africa where existing fixed-line infrastructure is limited. It is conceivable that with appropriate cellular policies on licensing, competition and pricing, cellular subscribers could eventually by-pass fixed link subscribers. For example, just two years after the introduction of cellular in Ghana, more than 3 per cent of all telephone subscribers are cellular users; with the current growth, cellular users would rise to a fifth of total subscribers by the turn of the century.

Table 5 : Cellular pioneers
Cellular subscribers in Africa, 1990-92

Economy	1990	1991	1992
Algeria	470	4'780	4'781
Egypt	4'000	5'240	6'944
Morocco	904	1'500	3'228
Tunisia	953	1'239	1'889
Gambia	0	0	204
Ghana	0	0	400
Kenya	0	0	1'100
Mauritius	2'200	2'556	2'912
Nigeria	0	0	*10'000*
South Africa	5'680	7'100	12'510
Africa	**14'207**	**22'416**	**43'968**
No. of countries with	6	6	10
No. of countries without	44	44	40

Source : ITU/BDT Telecommunication Indicator Database.

REGIONAL INTEGRATION

Though a number of regional political and telecommunication organisations exist in Africa, their synergy and effectiveness have been limited. One cause for the relative lack of success from regional co-operation may be that with an area so big and diverse it is unrealistic to expect that continent-wide regional groups can be very cohesive. Another reason is that historical links to European countries remain strong. African countries are literally not talking to each other and this is borne out by statistics on telecommunication traffic. If telephone traffic to and from South Africa is excluded, only 14 per cent of the traffic stays within the region, far lower than other regions.

Until there is strengthened inter-African trade, travel and tourism, the basis for regional co-operation in telecommunications is limited. Telecommunications has supported economic relationships in subregions where there is a significant degree of trade and travel like the southern African region. There has also been progress with continent-wide infrastructure projects but far more remains to be done..

Table 6 : Out of Africa
Outgoing international telephone traffic, selected countries, 1992

	Total traffic	of	Main	Bi-lateral	
Angola	9.1	6.6%	Portugal	5.1	55.9%
Côte d'Ivoire	21.4	27.2%	France	10.2	47.8%
Ghana	7'.i	7.2%	United	2.3	31.9%
TOTAL FOR COUNTRIES 18	513.5	38.2%		137.8	26.8%

Source : ITU/BDT Telecommunication Indicator Database.

PAN-AFRICAN TELECOMMUNICATIONS NETWORK (PANAFTEL)

The desire to install a Pan-African Telecommunications Network (PANAFTEL) had its conception at a meeting of telecommunication experts in 1962. Feasibility and pre-investment studies began in 1966 and implementation of the project began in 1975. By 1990, PANAFTEL radio-relay links in the region totalled some 38'000 km in length, 39 international telephone switching centres installed, 8'000 km of submarine cable laid and 42 out of 45 countries participating in the project had built international satellite earth stations.

Despite the fact that PANAFTEL has grown from a few isolated transmission links, its utilisation and potential revenue generation are not commensurate with the investment made and much telecommunications traffic that could be carried by the system is still being routed through transit centres outside Africa. Part of the problem stems from a lack

A regional network is dependent on each country doing its share. If one country does not maintain its portion of the network, then traffic flowing from other countries will be affected. The commitment of all countries needs to be assured to maximise the potential of the PANAFTEL network. A number of recommendations have been made for enhancing utilisation of the network including improving quality of service, completing missing sections, improving management, better traffic analysis, tariff review and proper marketing.

REGIONAL AFRICAN SATELLITE COMMUNICATION SYSTEM (RASCOM)

A couple regions -- Europe and the Arab-speaking world -- have their own telecommunication satellites as do a number of individual countries. In Africa, there is no regional or domestic satellite and the continent relies on INTELSAT and in some cases, ARABSAT, for satellite communications. This has drawbacks in terms of higher costs and technology which may not be appropriate in the African context.

The goal of the RASCOM project is to provide an efficient and economical means of telecommunications using appropriate technologies including a regional African satellite system. The system should be integrated into existing and planned networks and foster the socio-economic development of the African countries. The project has its roots in resolutions of the Conference of African Ministers of Transport, Communications and Planning and has the support of all main African political, economic and telecommunications agencies.

A project office was established at the ITU in 1987 to draw up a feasibility study. Country studies were made identifying requirements for satellite communications and the needs of rural and remote areas as well as regional integration. Technical and economic studies for the design, launching and operation of the satellite were completed as well as a specification of the equipment required and the possibility of local production. The project is now moving into the implementation phase. A headquarters has been established in Abidjan, Côte d'Ivoire, and is now operational.

The RASCOM feasibility study provided a unique opportunity for countries to examine their long term telecommunication requirements and in that respect alone was worthwhile. Furthermore, in part because of the collaboration on RASCOM, the African countries will be able to negotiate lower charges from INTELSAT by pooling their requirements for transponders.

POLICY OPTIONS

Africa is one of the last remaining havens of state ownership in the public telecommunication services sector. Only a handful of countries in Africa permit private sector ownership and this is either a legacy of former colonial relations (e.g., the role of Cable & Wireless, France Câble et Radio and Marconi of Portugal in various African countries) or has been introduced recently through the award of cellular radio licences (e.g., to Millicom in Ghana). In this sense, Africa contrasts strongly with the rest of the world where top 40 companies in the industry, which together account for over 90 per cent of total revenues, are dominantly privately-owned.

Africa also retains monopoly service provision to a much greater extent than other regions. Table 7, which is based on responses to an ITU/BDT questionnaire survey, illustrates that no country in Africa permits competition in voice services. Only a handful of countries permit competition in other services such as mobile, paging, data communications and satellite communications, and in some of these countries this is only theoretical competition as these services have not yet been licensed. In private networks the situation is a little better with around 50 per cent of the countries which responded allowing the use of private networks. Equipment markets are the most liberalised but here again, some countries, maintain state monopolies. The most liberalised countries in Africa, according to the survey, are Ghana, Morocco, Namibia and Senegal.

While some countries -- most notably in East and Southern Africa -- have established "public" corporations for the provision of telecommunications services, these are almost all combined with posts and with limited autonomy. In most of West and North Africa, services are still provided by government-run posts and telecommunication departments.

Several countries have partly "private" international telecommunication operators, generally co-owned by foreign operators with historical links. Whether they can all be considered private is debatable since some part owners -- France Télécom or Portugal's Marconi Comunicações Globais -- are themselves fully or partly government owned. In the former French colonies, the patterns is that there are separate domestic and international operators with the latter being minority-owned by France Télécom via its subsidiary France Câble et Radio. Only one operator in Africa is fully private and this is foreign owned: Cable & Wireless (Seychelles) Ltd. The situation is summarised in Table 8.

However, this situation is likely to change substantially in future years. The process of privatisation, which began in the Anglo-Saxon countries and has subsequently spread to Asia-Pacific, Latin American and Eastern Europe, is now coming also to Africa. A number of African countries, including Tanzania and Rwanda have announced definitive plans for privatisation and many other countries are considering it. Several policy options are available:

❏ Full-scale sale of shares in the operator to the public, to financial institutions and to employees either in one go or in several branches. This was the policy followed in the United Kingdom where BT was sold in three tranches between 1984 and 1993;

❏ Partial sale of shares in the operator through a public offering. This was the policy followed in Japan where just over around 35 per cent of NTT is now privately owned

.❏ Private sale to one or more investors, usually a combination of local investors and foreign owned PTOs. Sometimes the sale may be partial, as in Chile, Mexico and Argentina, and sometimes it may portend a sale to the public at a later date as in New Zealand.

❏ Introduction of private sector participation, for instance through a management contract (e.g. Cable & Wireless in Botswana) through the award of a mobile license (e.g. Millicom in Ghana). Other forms of private sector participation which have been used successfully elsewhere in the world but not yet extensively in Africa include the issue of Build-Operate-Transfer concessions (e.g., Shinawatra Group in Thailand), permitting local initiatives (e.g., community enterprises in Poland) or licensing competitors in the main fixed link network (e.g., Dacom for international services in the Republic of Korea).

Even if a policy of privatisation is not followed, it is still possible to give the PTO a greater degree of financial autonomy from the State through a process of corporatisation, or arms-length separation of the PTO from the State. In Senegal, for instance, the national operator SONATEL was separated from the corporatised and separated from postal operations on 1st October 1985.

Table 7 : Current state of market liberalisation in Africa, December 1993

	Customer premises					Voice services			Other services					
	Telep	PBX	Telex	Fax	Mod	Local	Natio	Inter	Mobi	Pagin	Data	Lease	VSA	Priva
Countries	33	33	33	33	27	32	33	33	21	18	29	32	31	28
Monopoly	10	5	17	1	4	32	33	33	15	13	23	30	27	14
Partial	7	5	3	4	2	0	0	0	1	1	3	2	1	7
Competiti	16	23	13	28	21	0	0	0	5	4	3	0	3	7

Note : The responses refer to the legal possibility of competition, not necessarily the actual status of the market.

Source : ITU/BDT Liberalisation survey

CONCLUSIONS

In most of the developing country regions of the world, there is optimism about the prospects for growth and development in the telecommunications sector, though the reasons vary. In Eastern and Central Europe, political change has acted as the trigger for a widescale restructuring of the sector and an opening up to foreign investment. In Latin America, economic stabilisation and a series of privatisations in the telecommunication services sector have generated new confidence and new growth. In Asia-Pacific, the underlying dynamism of the economy and openness to trade and cross-border investment is stimulating demand and spurring growth.

In Africa, none of these factors are relevant at a continent-wide level though they may be at work in particular countries. The 1980s was described as the lost decade for socio-economic development in Africa and the beginning of the 1990s has been hardly more memorable. The telecommunication sector illustrates this general malaise.

Africa's problem is not so much that it is disadvantaged relative to the industrialised countries, but rather that it is performing poorly in comparison to other developing country regions. There is increasing competition for investment funds, both from the private financial sector and from the multi-lateral development banks. The lack of good telecommunication projects in which to invest in Africa means that potential investors are favouring those regions of the world such as Eastern Europe, Latin America or South East Asia where there is a proven track record of growth and where there are openings for new market entrants. Telecommunications is also in competition with other, more high profile, sectors of the economy such as health care, education and water supply.

It is clear that the most pressing problems which the telecommunication sector in Africa faces are structural and *managerial*. There is a need to embrace private sector participation so that investment in telecommunications is not a burden on the finances of the State but is based upon a broader platform.

This, then is the challenge for African telecommunication policy-makers: to broaden the funding base for telecommunications development so as to better mobilise resources. The strategies adopted in different countries will vary according to whether the primary policy objective is to improve the performance of the incumbent public telecommunication operator or to introduce new market entrants. The menu of policy options may also vary between countries, but the basic ingredients are likely to be the same: private sector participation, competition, regulation, performance incentives, political commitment to telecommunications, accelerated investment programmes and regional co-operation. Above all, there is a need for regional role models which will show the successful application of these policy options in practice.

In other developing country regions, individual countries have taken a lead in triggering a wider scale sector restructuring. In Latin America, Chile was the first to undergo substantial reform and in Eastern and Central Europe, Hungary is currently leading the pace of change. Hopefully, in the next few years, a number of regional African role models will emerge which will have a similar continent-wide effect.

Table 8 : Foreign and private partners of
Public Telecommunication Operators in Africa, Dec. 1993

Country	Public Telecommunication Operator	Foreign/private partner
Botswana	Botswana Telecommunications Corporation (BTC)	100 % state-owned. Management contract with **Cable & Wireless**.
Central African Republic	SOCATEL	40% owned by **FCR**.
Chad	Société des télécommunications internationales du Tchad (TIT, international services only)	TIT is 43% owned by FCR.
Djibouti	Société Telecom International du Djibouti (STID, international services only)	STID is 25% owned by **FCR**.
Gabon	Télécommunications Internationales Gabonaises (TIG, international services only)	TIG is 39.2% owned by **FCR**.
Gambia	Gamtel	99% owned by the Ministry of Finance and Economic Affairs and 1% owned by Gambia National Insurance Company Ltd.
Guinea-Bissau	Guiné Telecom	51 % owned by **Marconi**.
Madagascar	Madagascar Telecom (international)	Madagascar Telecom 34% owned by **FCR**.
Mayotte	France Télécom	**France Télécom**.
Mali	Société des Télécommunications Internationales du Mali (TIM, international services only)	TIM is 35% owned by **FCR**.
Niger	Société des Télécommunications Internationales du Niger (STIN, international services only)	STIN is 23% owned by **FCR**.
Rwanda	RwandaTel	Created in 1993 from former government department. Invitations to tender for privatisation have been issued.
Sao Tome e Principe	Companhia Santomonese de Telecomunicações (CST)	51 % owned by **Marconi**.
Seychelles	Cable & Wireless (Seychelles) Ltd.	100% owned by **Cable & Wireless**.
Sierra Leone	Sierra Leone External Telecommunications Ltd (SLET, international services only)	SLET is 49% owned by **Cable & Wireless**.
Réunion	France Télécom	**France Télécom**.
Zambia	Posts and Telecommunications Corp. (PTC)	100% owned as a limited liability company by Zambia Industrial and Mining Company (ZIMCO) which is 100% state-owned.

Note : **Cable & Wireless** of the United Kingdom is fully private.
France Télécom, the parent company of **France Câble et Radio (FCR)** is fully state-owned by the Government of France.
Marconi Comunicaçoes Globais (formerly CPR Marconi) is 51% owned by the Government of Portugal.
Source : ITU/BDT Organisation Database.

Telecommunications and Development in Africa
B.A. Kiplagat and M.C.M. Werner, Eds.
IOS Press

CHAPTER III

AN ECONOMIC CONCEPT FOR UNIVERSAL SERVICE IN DEVELOPING COUNTRIES - EXPERIENCES IN CENTRAL AND EASTERN EUROPE

Odd Haugan
Senior Project Manager, Telecommunications,
European Bank for One Exchange Square
United Kingdom

INTRODUCTION

Since the first shopkeeper had his telephone installed, telecommunications have developed from a social instrument to a tool for the business community. The first shopkeeper probably saw the advantages in being reachable by his customers, and choose to invest in this novel instrument. I would be surprised if the investment did not pay for itself many times over.

Politicians, however, have in most developed countries and, in particular, in developing countries come to regard the social aspect of the telecommunications service as the more important aspect. Would this maybe have something to do with the voting power of the population in general rather than that of business communities?

Nevertheless, the telecommunications services we see today are equally revered by all that have learned to depend on them. The societies have developed habits that are irreversible, and can today not function without these services. The competitive edge lies in the availability of good information access, and billion of dollars are each year invested in more and more sophisticated telecommunications services.

According to the official experts, telecommunications is probably the fastest growing sector this decade.

AN ECONOMIC CONCEPT OF UNIVERSAL SERVICE

Universal service is largely a political dream. It will, in societies with the means to pay the investment cost, be almost a reality within this century, but it will certainly not be so globally.

Let us do a small summation of the cost involved. Assuming we look at the entire world population (say 5 billion people) with an average family size of 4 persons, we may define universal service as when every family has their own telephone connection, be it based on land lines, radio communications or satellite access. The requirement would then be, for residential subscribers, 1.25 billion telephone subscriptions. Assuming that there is a need for one telephone connection/subscriber line for each 20 inhabitants (using PABX's) in the form of business subscriptions, we have to add a further 250 million telephone connections/subscriber lines. We are now counting 1.5 billion lines world wide.

The question is, how many telephone subscriptions are in existence today? Let us for arguments sake assume that this number is in the region of 500 million. The need is therefore 1 billion telephone lines in total. Although development cost per telephone line can be as low as 800 US$ per subscriber lines in some low cost countries, let us assume that costs are rising in these countries and will eventually reach comparable cost levels of developed countries. If we then calculate the average cost of a telephone connection world wide, we may assume that the cost is around 1,500 US$ per subscriber line. The total investment cost is thus 1,500 billion US$ for world wide telephone service. The financial cost of borrowing such amounts of money will mean say 150 - 200 billion US$ per year. This, divided by the population served, being 1 billion, each have to contribute at least 150 US$ per year to pay for this investment. For a family possibly earning less than one US$ a day, this is absurd. Even for a family earning 10 US$ a day, the cost is excessive. Taking the acceptable proportion of telephone costs as a percentage of the total wage as being 1.5%, we arrive at a salary level requirement of 40 US$ per day to comfortably finance the cost of the telephone line investment. Not many households in Africa earns this kind of money.

Whether this figure is correct or not is immaterial. The point is, the absolute majority of these lines are required in countries of extensive poverty. Even if only 50% of the lines are required by the poor countries of the world, we still have an immensely difficult task in considering such investments.

The question for many a family will be, do they want a telephone, or do they want food. This is, sadly, for many millions of people the choice they must make.

Universal service, therefore, needs redefinition, or we need to abandon the political wish of such universal service. If we with universal service can consider access to telephone facilities when required, rather than the ownership of such facilities, we will see a totally different picture.

A Village Telephone Service (VTS) could satisfy the needs of a large group of people, who may pay according to usage of the service. If the investment cost can be spread on many users, the provision of universal access to telephones may be viable.

Taking into consideration the technologic development of satellite based services, such a concept becomes intriguing. Using electronic card technology[1] where this is relevant, and manned manual service where this is the preferred solution, one, or a few modern telephone sets of suitable technology can bring world wide access to virtually any remote region of the globe.

It is disappointing to see the world of telecommunications expertise discussing the rural telecommunications issue year after year without arriving at technical and service oriented solutions that could result in offering a better access to the broad populations. Again, obviously, strict market and turnover considerations are the dominating parameters rather than the social obligation to support the less fortunate and less affluent in the societies. A carefully balanced technologic development, a careful planning of the services required and provided and a careful selection of the financing mechanism can go a long way towards providing for those in need, but without the means. We are back to universal access, and not universal ownership as the basis for our consideration. Let us call this the Universal Access to Service (UAS), or Pay As You Use (PAYU). This is true Universal Service.

Now we may consider the economics of this form of universal service. The base criteria for service provision must be the ability to pay. No other criteria is workable in the long term. Even considering financing through taxes is not viable. The same people have to pay the same cost in the end, whether through their tax bill or through their subscription and call charges. There are no other sources of long term funding available. A little can be done using grants available to the poorest of nations, but assuming that grants to the

---

[1] The technology used in GSM mobile systems where the card defines the telephone number and not the telephone connections or telephone set. Any user owning a registered card can incert this in the telephone set, and the bill for the telephone call is sent to the card holder and not to the owner of the telephone connection.

tune of even a minor percentage of the required 1,500 billion US$ can ever be made available is wishful thinking.

We may, through careful engineering and careful choice of development strategy coupled with selection of financial strategy, obtain a favourable foundation for network development which in time could develop into a universal service for any country, assuming the national economy of the country in time would be able to support universal service. The bulk of the cost will, however, always be associated with the subscriber line plant or subscriber transmission system regardless of technologic development, since universal service envisage one connection for each household, therefore the long term investment cost for universal service can not be reduced substantially below 800-1,000 US$ per subscriber.

In Albania, with a present density of 1.45 telephone lines per head of population, and a monthly salary level of 30 US$, one should think that universal service is an impossible dream. One must, however, consider the economic potential of the country. With the second largest chromium deposit in the world, with oil production and oil industry, hydro electric power, several minerals being mined including copper, with a substantial agricultural potential and with ample tourist resorts just waiting to be developed, the potential for universal service in the future is clearly in existence. The choice of development strategy can therefore be based on getting services developed at lowest possible financing cost, and with technical and financial flexibility to adjust future development to demand development, while maintaining a profitable telecom sector.

The start in Albanian has therefore been based on a multitude of simultaneous efforts, starting with container exchanges provided as material grant by a benevolent telecom operator no longer in need of these exchanges. These provide up to 8.500 new subscriber lines, at a very low cost. The connection fee charged for these new subscribers help finance the first stage of the development programme, which will almost double the present penetration rate. This programme is financed by grants, soft loans and commercial loans. Some of the new exchanges will replace the container exchanges, which mean that the container exchanges can be moved to other areas and re-used there to provide "early" service to new subscribers. Eventually, the containers will end up in some rural area where it will provide satisfactory service for the community at a very low cost and for quite some time.

With such a substantial development at a low financing cost, the operator may find itself close to being able to finance subsequent developments by internal cash generation - or in other words - in a position of self financing. This makes the choice of financial strategy very attractive. Being able to afford to say no to a loan, certainly helps ensuring the cheapest possible financing in the future. The expected development in Albania is shown below.

Should the initial development have taken place according to traditional strategies, with a substantially larger initial development, including low profit rural networks, at higher financial cost, the outcome would be that the operator would be debt-ridden and unable in future to choose technical solutions and financing strategy.

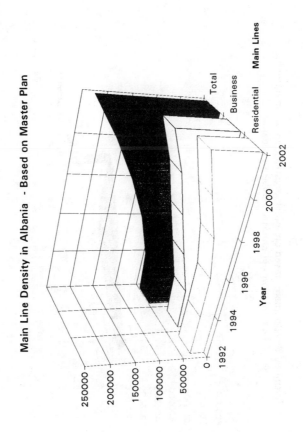

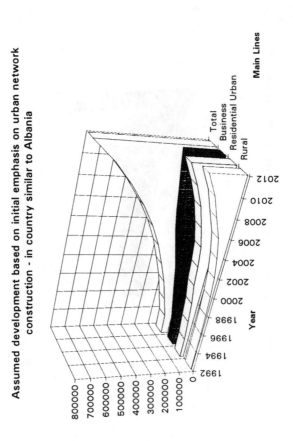

Assumed development based on initial emphasis on urban network construction - in country similar to Albania

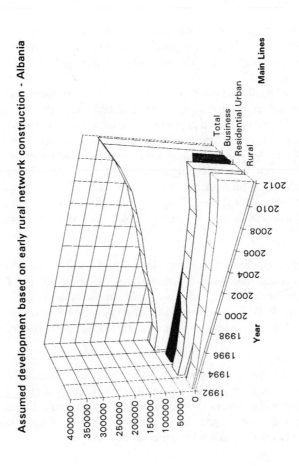

In the long term, the development strategy chosen for Albania, ensures better universality of development, and in the end quicker development of the rural network, although the first phase of development did not include the rural subscribers. The container exchanges will in a few years time prove a great blessing to rural subscribers, and in ten years time, the facit will show that development of the rural network has been more rapid than it would have been should the political wish be granted at the outset, to include the rural networks already from the start. Te comparison between the choosen strategy and a strategy including rural networks from the outset is indicated below.

Not all situations are comparable to Albania. If the national resources are severely limited, and the national product insufficient even in the future to support a comprehensive universal network, then alternative thinking is required.

So, let us therefore talk about Universal Access (UA) rather than universal service (US). Using VTS principle of development, the most suitable technological solution in each case to provide UA, we may see a potential economic solution to the lack of telephone facilities in the world wide village. Assuming that a poor village have a need for one telephone per 200 families and make one telephone call per month each lasting 5 minutes, the distribution between "local" and long distance calls being 20 to 1, we have 1000 call minutes a month, 50 of these being long distance calls. This translates to 12000 call minutes a year, 600 of these being long distance calls. This is a utilisation factor of approximately 7%, and is not too unlike a well used residential telephone in the developed world.

Assuming that the call charges are 0.1 US$ per 3 minutes for "local" calls, and 5 times this sum for the average long distance call, we have a total call charge of approximately US$ 480 per year for the connection. This would pay the interest of US$ 150 per year, the loan repayment of US$ 100 per year, and leave US$ 230 for operating costs. We are close to a viable UA service level provision for rural areas in the underdeveloped parts of the world.

Such investments may be economically viable, provide a satisfactory service to a population in need of such service and will satisfy realistic political ambitions rather than constitute an impossible dream. The continent of Africa could be covered by such services within a few years, and certainly before the end of the century, the most important bonus being the possibility to finance such projects.

PRIORITISING INFRASTRUCTURE ROLL OUT PROJECTS IN TERMS OF FINANCING REQUIREMENTS

Leaving the rural areas of our world communities, no comparison exist directly between urban and densely populated areas and the rural communities. With a satisfactory large population concentration and a substantial business environment, urban developments are almost always financially viable.

The suburban districts with less concentration of business subscribers may still be questionable investments, but similar concentration principle as those used for rural services may be devised.

A country is dependent for its welfare upon a business environment, which again is dependent on telecommunications services. To ensure that two vitally important aspects of telecommunications development is considered, the roll out of infrastructure projects have to be given a priority. Firstly, no network investment can be deemed viable without providing the large consumers of services with the means to generate the level of traffic required to finance the network. Building roads, at present and in the future, where no cars travel would be considered absurd except when there is clear indication that traffic will develop. Building telecommunications networks for the low traffic customers only must be regarded as equally absurd. Priority must be given at first to the high volume users. This is the essence of commercial orientation, to satisfy demand, first where the demand is largest.

But on the other hand, prioritising one customer group does not necessarily need to mean total neglect of others. A well balanced development can be constructed. Nevertheless, the revenues from the high volume traffic is an absolutely essential source of funds for investments in the less profitable portions of a telephone network. There is no discrimination in giving priority to the business subscribes and the initial preference then should not be seen as a promotion of the "elite".

We must not forget the social aspect of access from residential subscribers to business subscribers, remembering that a large number of residential subscribers may need to be able to telephone the local businesses that they would like or need to use. I would prefer to go to the telephone kiosk on the corner to phone the plumber, rather than having to take the buss to his workshop since he has not been prioritised with provision of a telephone connection, particularly when my cellar is full of water.

Arguments concerning technological development, privatisation, competition and foreign operators seem to miss one vital clue. The present public networks in all countries where such technologies, privatisation, competition and access to foreign operators are implemented have certain fundamental principles established that ensure satisfactory results from such adventurous undertakings as selecting these options entail. Certain countries have not established these fundamental principles, and when they try to utilise such solutions, the results are mostly disappointing and sometimes catastrophic. Without a basic modern telecommunications legislation, without free access to alternatives in respect of supply of equipment and services, financing alternatives and flexible tariff structures, without basic understanding in the population in general regarding competitive instruments and without consumer protection, these being some of the fundamental principles, the introduction of privatisation, competition and foreign operators can not safely take place without a high risk of abuse of technologic, financial and commercial dominance.

There is a good old British expression that we have to consider in this context: "Horses for courses". There are no patent solutions. Those that think that there may be such universal solutions available, are totally unaware of the problems. We should call these people "solution peddlers", for they have seen solutions and try to make them fit the problems. This clearly is nonsense in high technology and particularly in telecommunications.

Hence, the roll out of infrastructure projects is not always the right solution. A global development based on established statistics from developed countries is likely to generate a large financial burden in some countries without a corresponding revenue. The alternative may be an "island" approach, providing concentrated services where the ability to pay is established, and a minimum development approach (or VTS) where the ability to pay is reduced. For many countries in Africa, this will be lead to commercial centres being established with modern public network technology, and suburban and rural areas being provided with VTS, whilst desolate places will be provided with some form of satellite based direct access service operated by the small village "official".

It is well known that without an infrastructure in telecommunications, businesses will suffer from lack of interoperability and universal access to all subscribers. There is today no alternative to infrastructure development to create a universal network for telecommunications, but then again, at different times of the development of this infrastructure, one must take stock of the situation and see if there are better alternatives available. In essence, a long term plan is never right after it is implemented, therefore it must be reconsidered at each and every sensible opportunity, so that the future holds the most current thinking and not ideological relics of the past.

Infrastructure roll out, when that is right, should be considered as a basis for network development, but not at the expense of sensible alternatives when these materialise.

INTERPLAY BETWEEN SECTOR REGULATION AND LICENSING OF PRIVATE COMPANIES TAKING PART IN DEVELOPMENT

Without rules, no game will be played. Regulation of the sector may take many forms, depending on the political desire of the ruling authorities. Private initiative in the telecommunications sector will increasingly take the leading role in services development. If we can establish a sensible set of rules, and have the necessary authority to police the adherence to these rules, then the telecommunications market may be opened to all that have the financing will and ability to participate in the network and services development. To issue a licence does not constitute such a sensible set of rules. Licensing is but a tiny instrument in the large jigsaw puzzle that is covered by the term "sector regulation". Once the proper rules and the regulatory authority is established, a licence can be issued to any legally entitled juridical person that satisfy the requirements for being issued such a licence.

In Russia, a large number of licences for a variety of services have been issued. Unfortunately, the licences vary, and the licensees find the playing field rather uneven. Imagine the difficulties the authorities will have in ensuring compliance with more than a hundred licences, all different and not based on any single clear law.

The rules of the game must be of a positive character, or what we would call enabling rules, rather than disabling rules. The rules must state what the licence permits the operator to do, and only state the limitations to this permission, rather than stating purely what the operator is not permitted to. Otherwise we will find that the operator soon discovers loopholes.

The sector regulation must reflect the devised sector strategy which must be thoroughly debated and comprehensively approved by all major political parties involved in the present, and hopefully future, decision making within the sector. What one wants to avoid is political about turn at a forthcoming election, since investments in the sector by nature will be long term, and the investors want maximum political security.

Therefore, the role of the sector regulator and the status of a licence is all important aspects in attracting serious and long term foreign investments. The higher the security, the longer term the investment will take, and the lower the profit margin the investor will be willing to accept. The sector regulator is therefore not the right arm of the political authorities, but the adjudication body, as far separated from direct political influence as possible, and accountable preferably to a different state institution than that of the public operator.

The regulations and the licence terms shall clearly be of an enabling nature. The whole purpose of telecommunications legislation and regulation is to encourage service provision, reduce the cost of providing these services and, if possible, create a choice for the consumers of these services. Regulation is only a first step in the direction of commercialisation of the telecommunications sector. Once the supply of services exceeds demand and choice of suppliers of these services exists, the role of the regulator will be one of encouraging commercial enterprise and competition, and of preventing development of cartels and uncompetitive cooperation. In short, the regulator is there to prevent monopolisation.

The fear of the bypass operator is wholly unwarranted. The fear should be placed with the authorities and their inability to realise two simple rules of life:

❏ If a bypass operator can establish a profitable service based on a limited number of subscribers willing to pay for better service, then why can not the established public operator do the same? The investments are not very large, and the financing of bypass networks is available. The only stumbling block is the authorities and the regulations. In a truly competitive situation, the bypass operator and the public operator would have equal chance to obtain a bypass licence. The only reason this can not happen is the authorities preoccupation with the ideology of the universal service obligation. Why could the money from the bypass operation not be used to pay for new investments of less profitable nature, <u>regardless of who the operator is</u>?

Hence, the bypass operator would have an obligation to invest in less profitable ventures. All we need is sensible rules, and many things are possible.

❐ If a potential subscriber or an existing subscriber can not get through to the destination he wants to reach by any other means than travelling, then providing this subscriber with a bypass service until the public operator has developed satisfactory alternatives does not constitute a loss of revenues for the public operator. It constitutes, on the other hand, and increase in the national domestic product of the country. Only by giving universal access to the existing public network may the public operator suffer any loss of revenue. Therefore, the interconnect charges for such connections should reflect both the cost of provision of the service and such loss of revenues. Both the public operator would be happy, because he would earn increased revenues without having to invest in a bypass network, and the bypass operator would be happy, since he would earn increased revenues from a larger subscriber base without having to invest in the local network.

Bypass operations can therefore be directly beneficial to the public network development, and help to finance the development of less profitable services. Again, it is a matter of looking behind the surface and the well known "dangers", turning potential disadvantages into advantages for all.

Economics of Rural Network Development

Somewhere in the world, we shall always find someone who wants or needs telecommunications services, but is unable to obtain such services. Even with the satellite based services we envisage today, blind spots will exist. There will always be technologic reasons why extreme situations can not be covered, but we may limit these through development of technology to an absolute minimum (and should the technologist say that this is not true, then just consider someone without access to any convenient power source, we will not develop telecommunications running on candle power).

A much more serious limiting factor is the availability of capital and income to pay for the services offered. Without money, no telecommunications provider will be able to provide services. Looking at the capital needed and the access to development loans and private capital are just simply tools we can use in the development of services for those that can pay. If the subscriber can not pay, there is no way political ambition, cheap loans, private capital or foreign operators will make the slightest bit of difference. There will be no service provided to those that can not pay.

The rural solution is therefore closely linked to the economics of the proposed development. Providing services to a small rural community with the ability to pay and sufficient demand for services, will be a consideration of the traffic - and therefore revenues - generated versus the cost of providing the required services.

Rich rural communities with a high traffic demand have nothing to fear. Once the operator discovers the revenue potential, he will invest. The problem is the rural community with small demand and little ability to pay for the services. This is clearly the candidate for the Universal Access rather than Universal Service.

By limiting investment requirements and increasing traffic density, we can make a lot of people happy at a very low cost. In the rural community, people are used to have to move about a bit more than the urban dweller. Reasonable access to telephone service is so much more important than the convenience of this access. Not everybody needs a telephone in every room of their house to be satisfied.

Technology makes shared service more and more convenient. The concept of the card based identified subscriber, rather than the telephone apparatus based identification makes a lot of sense in the rural network. If we can link this to mobility and radio based subscriber connection or if needed to satellite based subscriber systems, we have a multitude of possibility, and can avoid having to string cables or wires to every farm in the land. Investment in the rural network should therefore concentrate on making services accessible in the rural community, and when the ability to pay is there, offer individual access to those that are willing to pay.

There is even a solution for the nomad, as long as he can pay for the equipment and the tariffs required to cover the cost of the service provided. But as even nomads travel in groups, every individual will not need a satellite based handset. Maybe one for the whole group is sufficient. The problem of billing the nomad is even not real, if one is willing to serve this market. Equip the set with a prepaid call card system, and the nomad can then only use his handset if he has already paid for the calls.

THE EFFECTS OF TELECOMMUNICATIONS ON OVERALL ECONOMIC ACTIVITY

Should we remove all telecommunications from the entire global village, we would soon find that the world and development came to a very abrupt standstill. Economic activity has from the beginning of time depended on communications. The bartering over a sack of grain against a spear or a few days work demands communications. Before telecommunications, we had to rely on messengers and later the postal services. The day the telegraph was introduced, our lives changed for ever. There is no way back. Today, no volume of transaction can take place without telecommunications. No society can today develop a competitive environment without the telephone, the fax or at least telex.

It has been seen in so many societies how supply of high quality telecommunications services has encouraged increased commercial activity, improved market access and made poor communities less poor, but the linking of telecommunications to economic growth is not entirely clear.

The question always arises about what comes first. Would extensive telecommunications development in Sahara crate economic growth? I believe we have to do what the telecommunications operators have done for so long, we have to put the horse before the cart, but ensure that it is the right cart.

We do have some live examples we can refer to. Telephone density in Hong Kong and in Singapore are certainly a pre-requisite for their extensive economic growth. In Ireland, the provision of efficient telecommunications services have enabled communities with few other opportunities to develop competitive data related services functions for firms located in the United Kingdom.

Lack of telecommunications is a serious hindrance. Provision of telecommunications enables economic growth, but does not create economic growth. Again, we are back to the essence of telecommunications development:

"Availability of the quality of services that the teleconsumers want at a price they are willing to pay".

When we link this with the essential challenge of service provision;

"Making available the services the teleconsumers want at a cost they can afford",

we see how the fundamental market forces are in active interplay. We see that supply and demand are as fundamental principles in telecommunications as in any other economic activity. Intervening in the demand development create artificial economy and may create a long term need for subsidies. This, of course, is a politically desirable intervention, as it may buy votes, but economically undesirable, as it creates a demand for services that are priced at below cost. The desire to increase tariffs to a cost based level is not there, as this will loose votes. The desire to provide cheaper services is not there, as the subsidies cover the cost in any rate.

The market forces must be allowed to work. Telecommunications is such an important pre-requisite for the market to function properly, that evaluation of the general market potential must be a foundation for the evaluation of the telecommunications market. Without commercial activity, the needs for telecommunications are limited to private and social requirements. We have a two tier system for evaluation of telecommunications demand, since there is clearly a two tier system in the usage of such services.

Similarly, we will have a two tier system in evaluation of commercial development, those without satisfactory telecommunications services and those with these services.

CONCLUSION

The African challenge is as large as the continent. Unimaginative attempts to solve the immense problem of telecommunications development in Africa without a comprehensive understanding of the real technologic, economic and financial long term implications is very unproductive, and may easily lead to the wrong solutions in the wrong places at the wrong cost and without the economic benefits resulting from the investments. Universal service is certainly not the right short or medium term goal for most of Africa.

Mistakes in telecommunications investments are very costly. It may in fact be so costly that mistakes in the past may prevent viable and much needed investments in the future. Unfortunately, the use of technically and financially highly qualified consultants is no guarantee for arriving at pragmatic and imaginative solutions. In fact, the problems of Africa are unique in character, and can not be solved using ready made solutions from the affluent countries of the Western World. Hence experience is not a prerequisite in itself, only experience in definition of problems and in devising imaginative solutions specifically to solve these problems.

What we require is lateral thinking and brave decisions taken by progressive authorities responsible for the sector and responsive to new thinking, but we do not want the solution peddlers. When someone comes to tell you that he has found the solution to your problem, the first question you ask him is: "Please, tell me what my problem is, then I may listen to your solution".

OoOoOoO

Telecommunications and Development in Africa
B.A. Kiplagat and M.C.M. Werner, Eds.
IOS Press

CHAPTER IV

AN ECONOMIC CONCEPT FOR UNIVERSAL SERVICE IN DEVELOPING COUNTRIES

Dimitri Ypsilanti
Head of Division Information Technology
OECD
France

INTRODUCTION

Among many policy makers from developing economies there is a misconception that telecommunication regulation is a luxury for countries with developed telecommunication infrastructures. This paper argues the contrary: effective telecommunication regulation can facilitate and accelerate the development process. In this context it should be clearly recognised that the goal of regulation is to achieve government objectives for the telecommunication sector, and those objectives of the national economy which telecommunications can affect. For economies with low telephone penetration rates a major objective is the rapid development and modernisation of the telecommunication infrastructure and services.

The African Continent, with perhaps several exceptions, is clearly lagging behind the rest of the World in the development of its telecommunication infrastructure (**table 1**). Even though the rate of infrastructure development, in terms of the number of main lines, has accelerated in the last decade, Africa still accounts for under two per cent of World main lines compared to a population share of 12%. The performance among the African countries in terms of infrastructure development is also highly variable reflecting, in part, weakness in growth of gross domestic product (**table 2**) but also insufficient priority at the national level to telecommunications.

Strict comparison of these data can provide a misleading picture of progress. Since the 1980s the network infrastructure in many developed economies has undergone profound change as a result of digitalisation. Quality of service has as a result improved significantly and new value added features have emerged in telephone service. As well, new networks have been put into place, notably for mobile services and data networks. This dual process of capital expansion and capital deepening implies that the real gap in terms of the availability and performance of the public telecommunication infrastructure between developed and developing countries is much greater than the statistical gap shown in Table 1. If account is taken of private networks, and value added network services in the developed economies then differences in the "telecommunications intensity" of economic activity between developed and developing economies is much more significant, and especially so between African and developed countries.

Table 1: Infrastructure Development: Number of Main Lines

	millions	millions	millions	Compound Annual Growth Rate			% of World Total		
	1971	1980	1990	1971-80	1980-90	1971-90	1971	1980	1990
OECD	150	250	360	602	361	474	9129	8819	8225
AFRICA	2	3	8	620	877	755	127	124	189
AMERICA	5	13	27	1047	718	873	335	469	613
ASIA	7	17	42	1080	973	1023	409	588	972

In Africa the infrastructure gap is not only a question of uneven development with the rest of the world, but uneven development between urban and rural areas. This problem is prevalent in many developing economies, and does not correspond to the economic characteristics of lower middle and low income countries where an important part of income generation is from rural areas. The difference in urban/rural tele-density is shown in Chart 1.

In many countries there has been increasing recognition that an efficient and developed telecommunication infrastructure can play an important role in economic and social development. Applying telecommunication to assist in the development of agriculture, natural resources, and manufacturing may facilitate improved self sufficiency and export growth. Moreover, telecommunications play an increasingly important role in unifying markets and, through demand and supply linkages, facilitating the efficient working of the price mechanism. This role is increasingly evident in production, sales, financing and management. Telecommunication also provides means to access national and international markets.

In this context, therefore, African countries must view the role of telecommunications as a tool to facilitate the economic development process and integrate domestic markets, as well as integrating domestic and international markets. At the same time telecommunications needs also to be viewed from the perspective of a social tool, integrating population, providing the means to aid effective governance, and the means to facilitate the provision of social services.

TELECOMMUNICATION POLICY AND REGULATION: ROLE AND OBJECTIVES

There is a large range of goals for the telecommunication sector **(table 3)**. For the development process three of these need to be emphasised: improving economic efficiency of the sector, creating conditions for security of investment, and promoting universal telecommunication service.

An efficient policy and regulatory framework, which is transparent and which minimises instability and uncertainty is a prerequisite for rapid development of the telecommunication sector. This is true irrespective of the type of market structure that is chosen. In the case of a monopoly it is important that the overall performance of the service provider is maximised such as through improved financial performance, enhanced economic efficiency, and sustained improvements in quality of service.

Given scarce investment resources, difficulties in access to foreign exchange and other investment constraints the requirement to promote efficiency is clearly necessary. There are a number of countries in Africa which, because of inefficiencies in procurement, have high investment costs per main line[2] or which have operating costs. High costs increase the necessary revenue per line requirements to generate sufficient revenues for further investment.

One of the requirements in improving economic efficiency in the telecommunication sector is to ensure that the tariff structure is compatible with the development goals of the sector. Telecommunications is a profitable business so that with the correct frameworks in place the sector can generate an important share of its investment requirements from internal resources, and support existing borrowings. An important source of investment funds is from the revenue of telecommunication operators which requires an efficient telecommunication tariff structure. This would allow sufficient revenue to partially self-finance development, support existing borrowings, and attract new investment.

---

2.   See, for example, World Development Report, ITU, 1994

**Table 2: Development of Gross Domestic Product per Capita and
Mainline Penetration Rates in selected African Countries, 1971-90**

	GDP per Capita (in US$)		CAGR (%)	Mainlines per 100 inhabitants		CAGR %
	1971	1990	1971-90	1971	1990	1971-90
Algeria	1287	2261	3.01	0.83	3.09	7.20
Benin	834	227	-6.62	0.15	0.31	4.06
Botswana	314	2340	11.16	0.44	2.04	8.45
Burkina Faso	252	335	1.51	0.04	**0.17**	7.79
Burundi	256	224	-0.70	0.07	0.19	5.66
Cameroon	972	586	-2.62	**0.16**	**0.34**	4.11
Cape Verde	731	**828**	0.66	0.49	**1.75**	6.97
Central African Rep.	316	423	1.55	**0.10**	**0.16**	2.70
Chad	792	167	-7.86	0.05	0.07	1.90
	1075	1317	1.08	0.50	0.70	1.75
Congo	1249	667	-3.25	0.40	0.59	2.14
Cote d'Ivoire	6964	**680**	-11.52	0.96	1.39	1.95
Djibouti	798	1200	2.17	0.89	3.25	7.05
Egypt	257	86	-5.60	0.14	0.25	3.09
Ethiopia	1954	3513	3.14	**0.50**	1.73	6.75
Gabon	477	243	-3.48	0.32	1.25	7.43
Gambia	627	342	-3.13	0.28	0.30	0.25
Ghana	260	247	-0.26	0.93	1.23	1.49
Guinea	364	303	-0.96	0.34	0.70	3.97
Kenya	210	**311**	2.07	0.14	**0.72**	9.18
Lesotho	235	192	-1.06	0.18	**0.24**	1.60
Madagascar	132	158	0.95	0.13	0.32	4.79
Malawi	627	177	-6.45	0.06	0.14	4.24
Mali	573	434	-1.45	**0.09**	0.31	6.79
Mauretania	698	1744	4.94	1.57	5.78	7.08
Mauritius	707	939	1.51	0.62	1.60	5.09
Morocco	**151**	127	-0.92	0.30	0.30	-0.05
Mozambique	711	**322**	-4.08	0.07	0.14	3.85
Niger	376	259	-1.94	0.14	**0.23**	2.51
Nigeria	**5802**	**8608**	2.10	1.88	26.80	15.00
Reunion	315	272	-0.77	0.04	0.14	7.28
Rwanda	877	493	-2.99	0.56	**2.01**	7.00
Soa Tome & Principe	1752	704	-4.69	**0.59**	0.60	0.10
	1864	5025	5.36	1.27	**11.52**	12.29
Senegal	416	127	-6.07	0.22	0.67	6.18
Seychelles	2109	2435	0.76	4.33	9.74	4.35
Sierra Leone	263	99	-5.01	0.21	0.26	0.95
South Africa	783	838	0.36	**0.78**	1.76	4.34
Sudan	213	114	-3.22	0.12	0.30	4.74
Swaziland	394	391	-0.04	0.17	0.30	3.10
Tanzania	601	1254	3.95	0.97	3.76	7.40
Togo	49	217	8.18	0.12	0.17	1.77
Tunesia	475	197	-4.52	0.11	**0.08**	-1.37
Uganda	447	440	-0.07	0.54	0.83	2.26
Zaire	477	585	1.08	**0.79**	1.34	2.79
Zambia						
Zimbabwe	613	609	-0.03	0.58	1.36	4.56
Total						

Notes: *Figures expressed in bold are OECD estimates;*
GDP per Capita is expressed in US$ at 1987 exchange rates and prices;
CAGR stands for Compound Annula Growth Rate.

Source: *ITU, IMF, UN.*

The countries of the OECD have accepted the principle that telecommunication tariffs should reflect both the structure and level of relevant costs. In most of these countries tariff rebalancing has been taking place over a period of several years. Only in countries with competitive telecommunication markets has the rebalancing process been much more rapid.

It would be naive to expect that countries in Africa could rapidly implement cost-oriented tariffs. It is, however, useful for policy-makers in African countries to require operators to begin this process. Cost-oriented pricing requires that operators obtain the necessary information to determine cost causation and put into place the necessary accounting and financial frameworks. This process in itself can help management improve efficiency, and provides regulators with information they require in policy implementation.

Underlying sector development there is the need to balance the issue of ensuring adequate investment resources for the operator against the requirements of industrial sectors which use telecommunications. In the early stages of network development both the considerations of the using sectors and universal service considerations should be secondary to the requirement to investment rapidly and expand the potential subscriber base. This means that the access charge (subscription) needs to be regarded in the early stage of development from two perspectives: it is a rationing device which can be used to manage (and set priorities) for the waiting list. It is also a means of obtaining scarce investment resources for further development of the network. Price discrimination is a valid tool in this context: charge high prices for those customers who are willing to pay to obtain a connection. Rather than being viewed as negative with regard to universal service considerations, such an initial policy will enable a more rapid network build-up and can be more effective in the medium term to facilitate the attainment of universal service.

The important **short-term** goal at the present stage of network development in most African countries is not universal service in terms of affordable access to the network, but to build-up the network as rapidly as possible. In Africa per capita incomes make the objective of an affordable **private** telephone service to the majority of households impossible given current costs; what is more important is that in each town, in each village the population at large have the possibility to have physical access to a telephone.

In the longer term, however, once a certain threshold has been attained in the subscriber base, revenue growth will then depend much more on usage. The above pricing structure is not necessarily incompatible with cost-oriented pricing since at the initial stages of development costs per main line tend to be much higher given that costs for switching and trunk lines are distributed over fewer customer main lines.

There are several lessons to be learnt from OECD tariff structures and their efforts to rebalance these structures. The first is the sensitivity (arguably misplaced) on upward adjustment of local call charges and access to the network. It is important that these charges are at least adjusted with inflation and operators (and regulators) are able to gradually bring these in line with costs. The second is that with the emergence of new telecommunication services (which did not exist a decade ago) cost-oriented access charges are important and therefore pressures to rebalance prices are today much stronger than they were when OECD countries were in the early stages of network development.

High capital costs of telecommunication investment, the planning and time required for effective network roll-out, and necessity for close ties with suppliers imply a certain stability in the conditions within which operators function. Governments need to provide such security and stability to operators. If countries wish to attract foreign capital to invest in the telecommunication sector then it is especially necessary that the appropriate frameworks are in place to ensure security of foreign investment. This requires at the level of the economy in general appropriate laws on foreign direct investment, and that government objectives for the telecommunication sector are clearly spelt out, preferably in legislation, and that the regulatory system is transparent and well understood by potential investors.

It is also important that the policy and regulatory framework protect the interests of users and other industry participants. This requires that the responsibilities of the public telecommunication operator, and the limits on its power, irrespective of whether or not it is a monopoly, are well established.

As competition is introduced in the telecommunications sector the function of the regulator to ensure fair competition between service providers becomes increasingly important. The transitional stage from monopoly to competition is characterised by the fact that the incumbent operator is in a dominant position because of a large customer base, access to customer information, an embedded network (often already amortised), and control over important resources (rights of way, spectrum resources, numbering resources).

The above arguments do not imply that universal telecommunication service is not an important goal. On the contrary universal service needs to be made explicit and should be based on specific targets in licenses which are provided to fixed-link operators. These should specify the coverage to be attained over a given time period in terms of penetration rates, the geographic distribution of this coverage, and penalties if these goals are not attained. What needs to be stressed is that at the early stages of network development, when there are very low penetration rates, the primary goal is to ensure that operators have adequate returns for their investment and sufficient revenue growth. This implies that business customers may have to be the first customer group to be served.

THE CHANGING INTERNATIONAL LANDSCAPE

It is commonly advocated that each country has the right to adopt unique national approaches toward the development of their telecommunication policy and to find their own solutions to these issues. However, it is increasingly difficult with increasing global interdependence for countries to consider change in isolation from each other.

Technological factors, the integration of economies and more open economic structures imply a requirement for greater consistency in policy frameworks of different countries. At the same time competition among economies for investment capital also implies a movement to greater similarity in underlying framework conditions.

Table 3: Policy Goals for Telecommunication Sector Development

- *Improve transparency of the conditions of telecommunications service provision*
- *Create conditions for security of investment and predictability of conditions for investors*
- *Enhance investment attractiveness in view of global competition for equity and loan resources*
- *Develop measures for encouraging regional co-operation to improve investment co-ordination*
- *Maximise cost reductions through increasing the internal efficiency of operators*
- *eliminate cross-subsidies between telecommunications and postal services, and unreasonable transfers from operator to government*
- *Clearly define the boundaries between monopoly and competitive supplied networks and services*
- *Improve the investment generating potential of tariffs*
- *Promote price liberalisation: rebalance tariffs while respecting social and economic conditions*
- *Clearly define the goals and representation of interest in regulation*
- *Clearly define the relationship of regulator with other government ministries and operators*
- *Achieve transparency of regulatory decision making including the licensing conditions of operators*
- *Promote measures to increase operator responsiveness to user demands*
- *Better define and enforce quality of service*
- *Encourage measures for enhancing operator efficiency*
- *Promote effective management of scarce resources such as numbering and frequency spectrum allocation*
- *Promote regulation that enhances technological development*

It is also important for developing economies to recognise that there are underlying structural shifts influencing the world economy. Economic activity in developed economies is increasingly based on service activities and manufacturing activities are slowly shifting to low-cost developing economies. The recently concluded Uruguay

Round enforces this trend, and may accelerate it. Among the developing economies there will certainly be competition to attract industries and, given the role telecommunications now plays in the manufacturing and distribution process, a modern network will play a key role in determining the relative attractiveness of countries. A number of South-East Asian economies have already recognised this and have taken steps to accelerate telecommunications development.

At the international level telecommunication networks and services are playing an integral role in the support and development of international market growth, and trade. These factors argue not only for greater emphasis on telecommunications development in investment priorities, but for greater similarity in market conditions and structures in which telecommunications activity takes place. Most countries in Africa, given the poor state of their infrastructure, can leap-frog a generation by investing in modern infrastructures.

REQUIREMENTS FOR DEVELOPMENT

It is an error to believe that developing countries can follow the models used by the industrial economies in the development of their telecommunication infrastructure. There are a number of reasons for this.

First, there are significant differences in the underlying economic and structural conditions between developing economies in Africa and the OECD countries in the mid-1970s. During the period when most OECD countries were developing their infrastructure their Gross Domestic Product per capita and and commensurate levels of disposable income were much higher than present levels for most African countries. Access to investment resources for developing countries are more constrained than they were in the past for industrial countries (government debts were smaller and public sector borrowing was less constrained, access to funds through financial instruments was easier). Hard currency was also less of a constraint than it is for many developing economies. The skill and management infrastructure in developing economies was available.

Second, however, there has been a radical change on the demand side as a result of the increased importance of telecommunications for manufacturing and services industries and for distribution including foreign trade. As well, demand has also become increasingly diversified with a wide range of sophisticated services being demanded by users, compared to earlier decades when telephony was the overwhelming service available.

The foregoing imply that there are a number of important differences in the financial and economic operating characteristics between public telecommunication operators in the OECD countries, in the 1970-80 period when a number of OECD countries rapidly expanded their infrastructure on the basis of monopoly market structures, and the PTOs of developing countries. For example, in the 1970s it was easier for OECD telecommunication operators to increase the relative intensity of investment using government financial guarantees. In developing economies with high debt this will be less feasible. Higher per capita incomes meant that the ability to generate investment funds from revenue was also higher. The data in table 4 show the ratio of telecommunications investment to total telecommunications revenue in the 1970s for some OECD countries. For developing economies it may be extremely difficult to raise such a high percentage of investment funds from subscriber revenue; ultimately it is this revenue which is important since loans need to be repaid.

It is certain, therefore, that without some significant structural reforms it would be extremely difficult for developing economies in Africa to obtain at the national level the required funding, or even a significant share of this funding, for investment. There are, however, a number of ways to generate investment funds:

❖ self generation of funds

This is clearly one of the key sources of funds. However, to be effective self-generation requires tariff restructuring, cost reductions from efficiency gains, implementation of new services and stimulating traffic growth.

❖ changing market structure

Monopoly market structures give limited opportunities for development, do not stimulate efficiency gains in the sector, and retard the implementation of efficient tariff structures. On the other hand in order to attract resources it may be necessary to provide an operator with some period of exclusivity. Hungary,for example, which has recently sold one-third of its monopoly operator to foreign investors provided an 8 year exclusivity period for international and national long distance telephone service and for over half of the designated local telephone areas. If such exclusivity is deemed as necessary then the regulator should attempt to provide performance incentives by other means such as through providing competition to fixed-link services from mobile communication services.

❖ foreign direct investment

An important source of investment funds for telecommunications can come from foreign direct investment. This could include the sale of a share of the operator to foreign interests, new market entry, and schemes such as "build, operate, transfer", etc. Many countries in the past were reluctant to allow foreign interests ownership in telecommunications viewing this area as a national resource and politically sensitive. Experience from OECD countries has shown that even without any ownership it is fairly easy for the regulator to provide appropriate control over the sector and the operator through regulation. The benefits a foreign investor brings through investment, technology, and operating know-how, far outweigh any hypothetical risks of foreign ownership.

Table 4: Telecommunications Investment as Per Cent of Total
Telecommunications Revenue

	1970	1975	1978	1980
Belgium	47.5%	60.6%	43.4%	42.3%
Germany	49.6	31.1	27.3	35.9
Italy	39.1	67.5	65.8	56.1
United Kingdom	48.8	49.2	25.3	31.6

MARKET AND DEVELOPMENT STRUCTURES

Developing countries need to address two issues which are separate but interlinked. First, what type of market structure is best to accelerate the development of the telecommunication infrastructure?   Second, where should priority be given in terms of the pattern of development of the national infrastructure?

In most OECD countries in the early stages of telecommunication development there was a belief that it was necessary to have a monopoly in order to develop the national infrastructure. There is no evidence in fact to show that a monopoly market structure was necessary. If the state of development of some major OECD countries is examined in the early 1970s the relative weakness in development is evident. In a number of cases only it was only through strong political action that a spurt of investment began.

In the process of changing to more liberal market structures for the provision of telecommunication services, whether this was for value-added network services, or for other services, the normal reaction among operators and some government policy makers in the OECD countries was to express concern that market opening would have negative effects. The type of effects envisaged included negative effects on universal service, and on the financial viability of the operator.   Experience from countries which have introduced facility and service competition   show that these fears have been incorrect. Universal service has not suffered from competition, rather as a result of lower prices, better quality of service, and through targeted programmes to assist appropriate low income households, universal service has improved.

The financial viability of incumbent operators has also improved through competition. Even though the former monopolies have lost market share as a result of new entrants, the size of the market has expanded with competition so that their total revenues has increased.

The market structure chosen as an alternative to a pure monopoly structure may have different implications upon the rate and pattern of telecommunication development and the nature and extent of regulation required. There are a number of alternatives to a pure monopoly structure. These can include: introducing a duopoly market structure, or allowing infrastructure competition; regionalising the country so that although a monopoly structure remains there is some "performance competition" between the different regional companies; a long distance and international monopoly may be retained and local service given to alternate operators. Whatever the choice it needs emphasising that competition can stimulate and speed development as different operators try to compete for customers, and competition will stimulate efficiency and prices which is to the advantage of customers and the economy.

There are also two possibilities that developing countries need to explore carefully: first is to allow small rural telephone companies to develop telephone service. Such small local companies have been successful and profitable in developed economies in sparsely populated areas. Their success in developing economies depends on allowing good management (preferably an experienced foreign joint venture) and following efficient tariff structures. The second is to allow several mobile operators to compete with the incumbent fixed link operator (and ensure that the fixed link operator is not involved in mobile service).

In most countries there has been a tendency to favour urban over rural areas in telecommunication investment. A monopoly structure will tend to favour such a bias whereas other forms of market structure will have different impacts on the urban/rural distribution of investment. The preference to rural telephony over urban is logical from several perspectives. First, an important part of national economic activity is concentrated in rural areas in many developing countries. Second, growth in rural areas will reduce urban pressures arising from population migration from rural to urban areas. A national operator may not have sufficient incentives to promote rural service since there are usually higher incomes to be earned from providing business urban services. It is for this reason that specific rural operators are preferable.

CONCLUSIONS

For developing countries it is important that telecommunication policies should be made in terms of an overall strategy to promote growth. They need to stress in their policy framework the close relationship between regulatory and institutional arrangements and infrastructure development. Policy makers need, **first**, to try to place sufficient emphasis on telecommunications and the need to stimulate telecommunications development at the highest political level. This applies mainly to the developing economies, but is also applicable to a handful of OECD economies. **Second**, efforts need to be made to augment the percentage of GDP that has been allocated to telecommunications investment. **Third**, more attention has to be paid to improve the efficiency and commercial viability of public telecommunication operators.

Clearly countries with limited financial resources will weigh investment in telecommunication against other investment. In this context it is worth noting that in general telecommunications investment yields high returns. In projects financed by the World Bank in Africa, Asia and Latin America the average rate of return on investment from telecommunications projects was in the region of 18 per cent. In an ITU/OECD joint study it was estimated that a 1 per cent rise in the number of telephones per 100 population over a five year period contributed in the 7 following years to a rise in average per capita income of 3 per cent. The contribution to national income was found to be particularly high in countries with the least telephone density.

Telecommunications also has beneficial social impacts. Telecommunications is also a profitable business so that, unless there is gross mismanagement, investment resources will lead to a sustainable sector which can generate an important share of its investment requirements from internal resources.

The requirement for effective regulation is not only to address the issue of growth and development but to address any concerns about the impact of competition. Even if these concerns were valid this should not lead to the conclusion that competition is undesirable, since regulation can in most cases help overcome any deficiencies that market mechanisms may introduce. But to prohibit competition would mean that the substantial benefits which competition has to offer in regard to enhance efficiency, quality of service, choice, innovation, diversity of products, etc., would be lost. As well, the ability to rapidly develop infrastructures would fade away.

The task of the regulator is not to manage the telecommunications sector, but to set down the appropriate conditions which allow the service providers in the sector to compete effectively, to perform efficiently, and to co-operate as required. The ultimate objective is the long-term development and performance of the industry. It is evident from the experience of OECD countries that as increased competition is allowed in telecommunications infrastructure provision and services, the role of the regulator increases, so that frequently regulatory requirements become more detailed and complex. This increase in regulation is not inconsistent with increased competition. Policy makers and regulators in developing countries need to prepare themselves to take on these new responsibilities.

OoOoOoO

Telecommunications and Development in Africa
B.A. Kiplagat and M.C.M. Werner, Eds.
IOS Press

CHAPTER V

EUROPE'S GREEN PAPER ON TELECOMMUNICATIONS AND ITS RELEVANCE FOR AFRICA

H. Ungerer and C. Berben (*)
Commission of the European Communities
Belgium

INTRODUCTION

This paper provides an insight into the development of a telecommunications regulatory policy for the twelve countries of the European Union (EU), probably the largest real life experience of supranational regulation yet undertaken. We feel that the underlying principles are relevant to the recent developments in Africa - not to be copied necessarily, but rather to be used as reference points from which regulatory scenarios can be examined, appreciated, developed and adapted to the realities of Africa. In much the same way, European regulators have benefited from experiences in other continents, and have learnt from the regulatory approaches of other countries.

The following key issues will be addressed:

❏ **The underlying regulatory principles, applied since the 1987 Green Paper, which now form the basis of the regulation of Telecommunications in Europe.**

❏   The consultative mechanisms, and the checks and balances chosen in a continuing process of regulatory development.

❏   The interrelationship of regulatory developments at the national, regional and global levels

The fundamental questions facing telecommunications in Africa are similar to those in Europe. What needs to be done to let the infrastructure serve as means to provide communications between persons and organisations? How can investments be attracted in a way which allows telecommunications to make optimal use of new technologies and market developments?

For the European Community, the new Treaty on European Union, which entered into force on November 1st 1993, amends the constitutional structure of the Community, and has - amongst other things - identified the establishment of trans-European networks as an agreed political goal.

It should also be mentioned that we have completed during 1993 a major re-examination of the telecommunications services sector, known as the "EC Telecoms Review". This process of review and consultation culminated in the decision to liberalise the voice telephony service - both at the local and at the cross-border level - by the 1st January 1998.

A consensus is thus emerging in Europe on the regulatory outlook. The agenda for the coming years has been set. There is broad agreement on its focus, based on the common convictions in the twelve Member States of the European Union and beyond, including the European Free Trade Area and the countries of Central and Eastern Europe. These newly-emerging democracies are tending to follow the European model in their market reforms.

THE MAIN DEVELOPMENT LINES IN EUROPEAN TELECOMMUNICATIONS REGULATION

The European Community's telecommunications regulatory process should be seen as part of the wider process of the political and economic integration of Europe embodied in the Treaty of Rome. This process was accelerated through the Community's internal market programme which since the mid-eighties has provided a firm basis for European telecommunications development up to 1992. From now on, this role will gradually be taken over by the broader political framework of the Maastricht Treaty, the Treaty on European Union. Within its framework, the European Commission, has shaped evolving telecommunications policy. In particular, the Commission under the terms of the Treaty enjoys the right to initiate policies for subsequent adoption by Member States in the Council of Ministers, after consideration by the European Parliament.

We feel that the stability of any supranational regulatory process will be critically dependent on its integration into a broader political and economic process - be it the Treaty on European Union, regional frameworks such as the OAU (or other arrangements in Africa), those offered by UN organisations such as the International Telecommunic-ations Union, or international multilateral trading agreements such as GATT.

The EU internal market programme has had its full impact on European telecommunications policy since 1987, the date of publication of the Green Paper on the EU telecommunications market[3], the first of a series of Green Papers or discussion documents published or planned for the sector. The major ingredients of the current process were:

☐ Introduction of a broad consultation process for policy formation based on Green Papers, the EU's consultative documents;

☐ Agreement on basic common principles of regulation: separation of regulation and operations; openness of procedures; dispute resolution mechanisms;

☐ Full use of the European Union's legislative powers to establish a policy of liberalisation and harmonisation.

AGREED REGULATORY PRINCIPLES

Liberalisation is progressively removing barriers to free operation across the EU's telecommunications market. The fragmented and disparate nature of each Member State's telecommunications market reinforced these barriers and as a consequence required measures of harmonisation to facilitate a level competitive playing for all market players. EU telecommunications policy has tried to reach a balance between liberalisation and harmonisation, since the consequences of too great an effort to standardise conditions of supply could well stifle the product and service innovations so vital to competition. The Major elements of this harmonisation have been:

☐ The European Community has set up procedures for the type approval of equipment for connection to a network in one EU country to be recognised universally across the twelve EU Member States. Similar proposals are being put forward on the mutual recognition of national licences to operate telecommunications services, in order to facilitate pan-European service provision;

☐ Extracting the full benefit from mutual recognition procedures depends critically on the development of common European standards. This was why ETSI, the European Telecommunications Standards Institute was set up early in the process

---

[3] *Green Paper on the Development of the Common Market for Telecommunications Services and Equipment (COM(87)290, 30.6.87)*

in 1988.   The GSM[1] European digital mobile system is one example of the success of strengthening of the standardisation process.

❐   Common principles for network regulation at a national level were agreed in the framework of Europe's Open Network Provision, ONP - a concept which has evolved in parallel to the US concepts on Open Network Architecture.   The regulatory principles being developed aim at harmonising the conditions under which users and service providers can access telecommunications services and facilities across the European Union and the minimum levels of service which should be provided.

The Open Network Provision programme has developed into a broad framework for the development of future regulatory principles for the supranational process in Europe. During the 1992 Review, a consensus emerged such that ONP legislation - both adopted and in process - constitutes the right basic framework for future network regulation at the EU level.

These basic principles develop along three lines, all of them included in the ONP Framework and associated specific Directives[2], in particular those covering voice telephony and leased lines.   They concern: universal service; interconnection rights and dispute resolution mechanisms as they relate to both users and operators.   All of them are also major topics on the international agenda.

The concept which without doubt has had the deepest impact on telecommunications reform across Europe has been the principle of separation of regulation from operations. It led to a deep transformation of the industry (often previously under the direct control of the national ministry) and laid the very basis of building an effective supranational regulatory process.

At the same time, it is leading to an arms-length relationship between the sector and the state and to substantially more flexibility for the telecommunications operators to adapt to the new environment.   It is this new flexibility which has led the traditional operators to largely embrace the regulatory reform - a major prerequisite for the success of the process.

---

[1] *GSM - Global System for Mobile communications*

[2] *See*
Directive 90/387/EEC, OJ L192/1, 24.7.90 (ONP Framework Directive),
Directive 92/44/EEC, OJ L165/27, 19.6.92 (ONP for Leased Lines),
Proposal on Voice Telephony, in process of adoption.

CONSULTATIVE MECHANISMS

A broad consultation process - based on consultative documents, Green Papers - has played a key role in the process of supranational market reform which the EU has initiated. It has been an essential element in consensus-building amongst political partners, operators and all market participants which must form the basis of any supranational process.

The European Commission has consulted extensively with all sections of the industry and users across the twelve Member States. Let us illustrate this by giving some detail on the Review of the Telecommunications Sector mentioned earlier. This review process was initiated in October 1992[1], with the publication of a policy paper on the development of the telecommunications sector in the European Community, setting out the basic policy options for the liberalisation of the voice sector, ranging from maintaining the *status quo* to the full liberalisation of voice telephony services.

Views were welcomed on these general policy options and on the direction of future policy. Indeed, the contributions received led to a re-focussing of the original approach (centred on opening inter-exchange competition) to one advocating full scale liberalisation with a transition period for its implementation lasting up to 1998.

By January the consultation process was largely completed. Many written submissions were received. These included the views of a broad spectrum of user organisations, commercial enterprises, telecommunications operators, service providers, equipment manufacturers and trade unions not only across the twelve EU Member States, but also from the rest of Europe.

In parallel, a series of public hearings took place. Both the European Parliament and the Economic and Social Committee were consulted.

During the whole process, the Commission worked closely with a high-level committee of the heads of the national telecommunications regulatory authorities. The Member State regulators were involved both in the structuring of the public hearings and the evaluation of the results. They reviewed the final report[2] and recommendations produced by the Commission.

This example is illustrative of the consultation process which is required in the supranational process. Extensive consultation of all interest groups and transparency in carrying out the process is essential to achieve stable market reform in a multinational environment.

---

[1] *Communication of 21st October 1992 on the 1992 Review of the Situation in the Telecommunications Sector (SEC(92)1048).*

[2] *Communication of 28th April 1993 on the Consultation on the Review of the Situation in the Telecommunications Sector, (COM(93)159).*

THE RELATIONSHIP BETWEEN REGULATION AT MEMBER STATE AND SUPRANATIONAL LEVELS

A major issue throughout the development of telecommunications policy within the EU has been the balance between regulatory oversight at a European level and its relationship with the national regulator.

The European Union acts in conformity with was is called the **principle of subsidiarity** which attempts to draw the limits between Community action and national action, in those areas where the Community enjoys specific powers under its founding Treaties. According to the Treaty on European Union, now in force, the principle of subsidiarity means that "*the Community shall take action only if and in so far as the objectives of the proposed action cannot be sufficiently achieved by the Member States and can therefore, by reason of the scale or effects of the proposed action, be better achieved by the Community*".[1]

In consequence, the Commission envisages taking the "*minimum action necessary at Community level in order to remove obstacles to the provision of the widest possible range of telecommunications services. Within the framework thus created at Community level, Member States will continue to determine their own telecommunications policies*"[2]

This principle is incorporated in the EU's basic regulatory framework, whereby the Member State telecommunications regulatory authorities have the primary responsibility for regulation at the national level. The role of the EU's supranational regulatory process is mainly to spell out general principles, to offer conciliation in disputes which cannot be resolved at a national level and to assist in establishing a consistent approach throughout the Union.

It is illustrative to look at the example of future regulation of **interconnection issues** as envisaged under the ONP framework. Responsibilities are to be shared as follows :

At Member State level:

❐   the national regulator has an obligation to intervene if requested by either party ;

❐   the national regulator has an obligation to ensure that interconnection agreements are non-discriminatory, fair and reasonable to both parties and offer the greatest benefit to all users;

❐   the national regulator has a right to impose conditions in interconnection agreements concerning technical standards, quality requirements, conditions safeguarding so-

---

[1] *EC Treaty, Article 3b*

[2] *Communication (SEC(93)1048), note [3]*

called essential requirements, and to impose deadlines for implementing
interconnection agreements;

❑ Member States must establish procedures for resolving disputes between
telecommunications organisations and users.

At the level of the European Union:

❑ users may invoke a conciliation procedure to resolve disputes not resolved at a
national level;

❑ the Commission may take measures to secure the provision of particular pan-
European services.

The European Commission may also invoke the competition rules in the case of abusive
behaviour or anti-competitive agreements, as it has done in a series of cases in the past.

AN EXAMPLE OF SUBSIDIARITY - THE ISSUE OF PRIVATISATION

Both the Treaty on European Union and a strong respect of the national right to
determine the organisation of the national telecommunications sector set strict limits to
supranational regulation in Europe. One major issue falling under this category is the
issue of privatisation, so much so that privatisations in the telecommunications sector will
move to the top of the policy agenda in Europe during the coming years. Recent reports
have estimated that in Western Europe alone, telecommunications privatisations may
account for three fourths of the total value of all privatisations which may take place.

Article 222 of the Treaty provides that the Community shall *"in no way prejudice the
system of property ownership"* in Member States. Therefore, the Treaty prescribes that
the issue of privatisation or maintenance of public ownership falls within the national area
of competence and is not subject to policy-making at the European level.

However, the fact remains that the issue of finance and the possibility of contributions
from private capital is faced by all operators across Europe. The full-scale liberalisation
schedule for voice services now set for 1st January 1998 has further accelerated this
movement.

Throughout Europe, telecommunications operators are converting to shared ownership
or to publicly-quoted companies. The third and final tranche of British Telecom has
reached the market. Plans for partial privatisation of France Télécom have been
discussed and similar plans for Deutsche Telekom are in train in Germany. There are
privatisation plans in the Netherlands, Italy and Portugal.

Requirements are, of course, particularly pressing in Central and Eastern Europe, where in the context of general market reform, **enormous investment is needed** to modernize telecommunications infrastructure. The massive long-term capital base required for network build-up will inevitably require large scale injection of private capital, as current developments in Hungary and the Czech Republic indicate.

These financial requirements will be a dominant element in developing telecommunications reform during the decade. For the developing countries and for regions such as Central and Eastern Europe, to a substantial extent the international development banks set the conditions for such financing and assist in decisions on privatisations. This makes them important new actors in scrutinising and developing regulation. Reference is made here to the activities of the World Bank, as well as to regional financial institutions, such as the European Investment Bank, the European Bank for Reconstruction and Development and the African Development Bank.

SATELLITE COMMUNICATIONS

Satellites play a vital role in the provision of services in those regions which cannot be serviced optimally by terrestrial infrastructure. In regions like Africa, new possibilities are arising for the development of telecommunications networks and services. It is therefore essential that rapid progress be made to permit optimal use of the new possibilities offered by technology. In particular the recent developments in VSAT technology (very small aperture terminals) has brought small dishes (of the order of 1 meter for two-way communications) to the market, at prices which are relatively low.

In Europe, this has resulted in the publication of the Green Paper on Satellite Communications in November 1990, and a subsequent open consultation. The document aims at a fundamental restructuring and liberalisation of this sector.

Four strategic objectives have been formulated:

❐ Liberalisation of the earth segment, including the abolition of all monopoly rights; use without restrictions of receive-only equipment ; liberalisation of two-way communications equipment, subject to type approval and the issuing of a license, in particular to avoid interference problems and to ensure frequency coordination.

❐ Access to the space segment without restrictions, subject to the issuing of a license.

❐ Service providers will have the possibility to obtain space segment capacity through direct negotiations with satellite operators. This implies commercial freedom for space segment operators like INTELSAT, INMARSAT, EUTELSAT to offer capacity directly to users and service providers.

❐ Harmonisation measures, where necessary to promote the use of standards to facilitate agreements on type approval of equipment, to coordinate frequency use and to solve the questions linked to service provision across borders.

Currently in Europe, the above proposals for the satellite sector described above have either been adopted (type approval of terminal equipment), or are under examination (mutual recognition of licences, access to the space segment).

COORDINATION OF FREQUENCIES AND NUMBERS

The progress made in the field of mobile communications requires a coordinated approach at the European level in order to respond to the regulatory issues which have arisen. The major activities of the European Commission in the area of mobile communications have focused on the reservation of frequency bands for digital mobile systems: GSM, DECT[1] and ERMES[2]. In addition, the production and use of international standards has been promoted in order to ensure European-wide interoperability of services.

A European Radio Communications Office (ERO) has been established in Copenhagen, with the task of making the coordination of frequency allocations more efficient and the process more open through consultation with users, industry and service providers.

In a similar manner, the fair and non-discriminatory allocation of numbering capacity has become a major element for the efficient control of telecommunications and the creation of an environment of competing provider. In Europe, as in Africa, the management of numbering plans will become a major regulatory task - both at the national and at the supranational level.

The Mobile Green Paper, which is to be published in the early spring of 1994, will address a broad range of issues including frequency and numbering allocation, and thus invite an open discussion at the European level as to how to develop this promising sector.

THE GLOBAL CONTEXT

The ultimate test for the regulatory framework in Europe is whether it can stand up to scrutiny in the wider global context.

Europe's historic framework of contact with the international telecommunications environment has been the International Telecommunications Union (ITU). With progress in liberalisation and national regulation, the negotiations on the General Agreement on Tariffs and Trade (GATT) have become an additional focus of attention.

---

[1] *DECT - Digital European Cordless Telephone.*

[2] *ERMES - European Radio Messaging System, the new European digital paging system.*

The change in the international regulatory environment is leading to fundamental challenges for the international regulatory scene and for the ITU. In response, the ITU's High Level Committee has initiated an overhaul of the organisation which has given it a more efficient organisational base in standards-making, coordination of radio frequencies and satellite orbits, and for its action for the developing countries.

However, the main effects of current changes at the national and regional levels on the international regulatory scene have yet to be felt:

❐ the liberalisation of the major part of the world's telecommunications markets is currently leading to a rapid erosion of the existing accounting rate system and a redistribution of financial revenues which may lead to a substantial weakening of the telecommunications financial base of the developing countries. Such countries represent a major part of the ITU membership and this will inevitably pose a major challenge for international coordination;

❐ the fundamental changes in the international balance of telecommunications between regions in the developed world, as well as between the developed world and the developing countries, are difficult to accommodate within the existing international telecommunications framework, and are likely to lead to substantial tension. This concerns both the International Telecommunications Union and the international satellite organisations such as INTELSAT;

❐ at the same time, new global systems are developing which have a substantial potential for bypass of the existing structures both in technical and regulatory terms. The major immediate issue on the agenda in this respect are the new satellite-based personal communications systems such as those discussed in the context of Low Earth Orbit satellite systems, LEOs. Given that structures to establish supranational regulation for these new systems are not currently in place, national regulatory decisions taken in a hasty manner will inevitably tend to prejudge - or even to replace - appropriate the supranational regulation which must develop. The current US FCC licensing process for LEOs is a case in point.

❐ the new commercial aspects of the liberalised telecommunications equipment and services trade cannot be easily accommodated in the current framework by the traditional mechanisms of international telecommunications coordination. This necessarily establishes GATT as a major new forum for international telecommunications regulatory issues.

In the process of the GATT negotiations, the inclusion of trade in services - the GATS round - marked a major step in the scope of the negotiations, reflecting the world-wide maturing of the services sector. The supranational regulatory process in Europe - and particularly the Open Network Provision concept - provided substantial input into the negotiating package.

CONCLUSION : EUROPE AS A PARTNER IN THE DEVELOPMENT OF TELECOMMUNICATIONS IN AFRICA

The major challenge of this decade for the telecommunications sector will be to define its role in the development of the world-wide economy.

The relationship between telecommunications and economic development has been the subject of numerous publications, and in this context we refer only to the recent World Telecommunications Development Report 1994, prepared for the ITU. The social impact of telecommunications may be more difficult to measure, but is equally significant. The impact on safety, accessibility of peripheral regions and development is evident. It is therefore no surprise that the concept of Universal Service draws the attention of regulators across the world. The concept needs to be defined in the context of each continent and each country, based on the resources available.

It is our view that telecommunications can and must play a vital role in the development of any society. To do this successfully, conditions need to be created which encourage capital investment, the participation of entrepreneurs and the optimal use of scarce resources. The regulatory environment can contribute by providing transparent, stable and non-discriminatory access to telecommunication resources.

A large responsibility is put on the regulatory authorities to design the instruments which allow society as a whole to benefit from new technologies and services which have proven their viability. Competition can be a major driving force for innovation, attraction of investment funds and customer satisfaction. Of course, the formulation of public interest goals, for example those regarding to universal service, are indispensable.

We are convinced that these public interest goals can be met in a liberalised environment. The issuing of licenses is an appropriate instrument to ensure that public service requirements are met and shared in a fair way between the parties involved.

Such an environment permits parties to develop their activities with a minimum of interference, while the regulatory authority keeps the reserve powers to intervene when required. Such an environment also invites partnerships, between national operators on the one hand, and investors, manufacturers and international service providers on the other.

This applies equally to Europe and to Africa, where huge investments in infrastructures are needed to benefit from modern telecommunications. The uniqueness of the European model resides in the balance applied in restructuring the sector.

It is essential that open access to networks and services be ensured, while preserving the public service goals and the national identity of each country.

Every country must find its own model for development, respecting fully its own national identity. In this context, the European Commission is prepared, in bilateral or multilateral contacts, to share its regulatory experiences in the field of telecommunications, in addition to the substantial development effort which it already undertakes in partnership with the countries of the African continent.

OoOoOoO

(*) Herbert Ungerer is responsible for regulatory aspects of telecommunications network access, satellite communications, mobile communications and frequency management in the Telecommunications Directorate of the European Commission. In this capacity, he was responsible for the preparation of the 1987 Green Paper on European Telecommunications, and subsequently for the Green Papers on Satellite and Mobile Communications.

Cor Berben is responsible for Access to Networks in the Telecommunications Directorate of the European Commission, and is in charge of the ONP programme, aimed at the availability of European-wide telecommunications networks and services, as well as the open access to and utilisation of these networks and services. He has also participated in the discussions on open access in the multilateral platforms of the GATT and of the EEA.

The opinions expressed in this paper are solely those of the authors.

Telecommunications and Development in Africa
B.A. Kiplagat and M.C.M. Werner, Eds.
IOS Press

CHAPTER VI

"THE AFRICAN GREEN PAPER" AND ITS HOME AUDIENCE

Etienne Konan Kouadio
Deputy Director, Legislation and Regulation
Ministry of Transports and Telecommunications
Côte d'Ivoire

The initiators and the editorial committee of "The African Green Paper" (1) deserve our sincere congratulations, for the ambitious project they have undertaken and the quality they achieved. We would encourage them to continue their work, reexamining a number of aspects which in our view have not been sufficiently addressed or clarified.

These observations, intended as modest contributions and improvements of the "green paper" development process, reflect situations which can be encountered all over Africa and touch upon current questions and topics which may look trivial but nevertheless deserve our consideration. Topics in the "green paper" which we consider sufficiently covered are given less consideration in this paper.

With regard to the name "The African Green Paper", we all know that in 1987, the "Green Paper on the Development of the Common Market for Telecommunications Services and Equipment" was launched at a European Community level. Likewise, in 1990 the "Blue Book on Telecommunications Policy for the Americas" was published, targeted at the United States. For Africa, a project entitled "Telecommunications Policies for Africa - The African Green Paper" is under preparation. If this title were to be maintained, it would appear slightly embarassing to see two publications under the same title of 'Green Paper'. Since in common language one would simply refer to 'the green paper', this situation may lead to lead to confusion. Besides, suppose that Europe has chosen the color green to symbolise its green pastures, and the Americas the color blue to symbolise the oceans, why should Africa not select orange or ocre to symbolise its savannas? We therefore would like to propose that instead of The African Green Paper the title should be The African Orange Paper.

A POLITICAL FRAMEWORK

"Some operating agencies do not even have the capacity to deploy efficiently the financial resources they would ... need ... This traditional approach based on State-owned enterprises has shown its limitations" (o.c. para's 10 - 11). However, it is not always fair to continuously blame the traditional approach for unsuccesful use of financial resources. What would one say about all those private firms that are forced to close shop following bankruptcy? This is certainly not enough reason to dismiss private enterprise. We would rather like to maintain that this incapacity be attributed to incompetence, poor management and wrong management decisions. We propose three solutions to this problem:

(1) Governments as (majority) shareholders should not infringe upon the management responsibilities of the operating entity, particularly with respect to recovering debt;

(2) A reasonable and adequate structure for controlling staff, including application of staff regulations;

(3) If in spite of the above, irregularities are committed by staff and management, sanctions should apply, however, without animosity nor vengeance.

In line with these points, a new policy for the creation of a dynamic and consistent policy framework is called for, in order to realise the development objectives of the telecommunication sector. With regard to a political framework it is important to recognise that one cannot formulate policy goals without establishing the associated policy measures and financial resources. It is also important to ensure that policy goals are well prioritized and evaluated against technical expertise. Even if it is the government which is setting such policy goals, this should not preclude participation of technical experts in the decision making process.

THE LACK OF INFORMATION

The importance of information in a corporation or governmental organisation merits our particular consideration.

One aspect pertaining to the African environment which has our prime attention concerns organisational, rather than technical or even political and legal realities. We are referring to a level which is situated much lower than that which is addressed in the "green paper". We are referring to information and the dissemination of information.

The retention of information inside corporations and governmental organisations is commonly practiced and we are aware of the reasons thereof. Why is information retained? To mislead? Information related to policy making does not flow as it should. It flows poorly and leaves gaps. As a result, staff which has a need to be informed in order to be efficient and perform well, cannot gain access to crucial sources. Those who retain information hold on to it and manipulate its flow. Sometimes they make it available to the right persons, but sometimes also they guard it for their own benefits. The way in which vacancies in the ITU for instance or regional organisations, and even in national agencies, are announced is illustrative of this practice. Although such vacancies should be broadly advertised in order to encourage the best candidates to come forward, they are hardly ever brought to the attention of staff who are often very well qualified. The same applies to journals and periodicals of relevance to, in our case, telecommunications. Those are important for staff to update their knowledge on technology and regulation, but are rarely made sufficiently accessible. Other examples concern information about scholarships, trainee positions, seminars and conferences. The above examples indicate clearly that the correct circulation of information is mandatory, both horizontally and vertically throughout an organisation.

SUSTAINABILITY OF NETWORKS AND SERVICES

It is commoly recognized that Africa has the least developed telecommunications network in the world, for reasons well known to us and enumerated by the "green paper": "The infrastructure capacity is insufficient and most infrastructure equipment comes from suppliers outside the region. Moreover, the equipment is produced according to standards and at a rate which takes no account of Africa's specific needs and environment. Most of the time, there is a great diversity of equipment within the networks ... Network and service quality are far behind prevailing standards and network operation is conducted by over-abundant and generally little motivated staff" (o.c. para 54).

Careful analysis of the causes of underdevelopment allows us to categorise these in three categories: the infrastructures, the African environment, and the management and operation of the networks. Which explanations could we offer?

The majority, if not all of the telecommunications infrastructures of Africa remain public assets, and their inefficiency could be explained by the fact that Government administrations do not always have sufficient means to expand them, as they are confronted with many other imperative tasks, such as improving public health, education and road networks. From our point of view, it would be worth drawing the attention of the public authorities once again to scientific evidence of the high economic benefits of telecommunications and urge them to allocate a more important share of the investment resources they control. Meanwhile however, it should be well understood that once built, these networks require continuous maintenance in order to prevent them from degenerating into a state of disrepair.

Given the costs of equipment for telecommunications infrastructures, it is important to note that such equipment should only be ordered after thorough evaluation and identifying a well focused application area. In addition, the diversity of such equipment requires that only those which are best suited for the intended environment are selected, rather than acquiring at any cost those which are offered by the traditional suppliers from countries which are preferred for political reasons.

It seems particularly important to note, considering that equipment is designed and produced outside and primarily for non-African countries, Africa should keep in mind its own requirements and rythm, rather than live at the rythm of suppliers and acquire as a matter of routine and at any cost, equipment of the latest generation, which is often unsuitable or not adaptable for our network configurations. This truism is highly regrettable and hurts our interests. A closer look inside certain telecommunications premises within our region allows us to see expensive stocks and equipment, acquired a long time ago but gathering dust in warehouses. Meanwhile, these materials and equipments require proper maintenance: appropriate storage, expert installation, rational utilisation and permanent protection. Disregard of such measures will lead to premature degradation of stocks and installations, as has been the case in a number of countries.

In the absence of an indigenous African telecommunications industry, we would recommend a structure for the creation of regional standards, which would define and elaborate technical specifications for equipment, taking into account the environmental conditions prevailing in Africa. This would be far more efficient and beneficial than trying to create such a structure in individual countries. Furthermore, it is unclear whether each country would be able to set up such a structure, for lack of funds and qualified personnel.

HUMAN RESOURCES AND NETWORK OPERATION

The functioning of a telecommunications network can only be optimised with the full commitment of qualified technical and management staff, regardless of the level of network digitalisation. Certain working conditions must be met to ensure that staff can fulfill its role: proper work premises, suitable tools and support systems, proper remuneration including bonuses, as well as continuing education.

At this point it is relevant to note that, since independence was achieved following colonial rule, the African telecommunications operators have created large staffs. Unfortunately, many well educated and qualified staff members are being frustrated and their skills under-utilised. They are often even assigned duties that have no bearing whatsoever with their professional background. The obvious consequences are the following:

☐ he state is the first loser, since it gets no returns out of its efforts in providing professional education;

☐ he operating agency as employer is the second loser, since it is paying salaries without utilising available knowhow;

☐ he employee is the third loser, since he has to adapt to new areas of duty while losing previously acquired expertise.

We recommend that efficient human resource policies be developed, in a dialogue between the relevant government authorities and decision makers within the operating companies. The objectives and costs of human resource development and permanent education should be clearly defined. It is evident that poorly or untrained personnel is in no position to deliver the telecommunications services for which they carry responsibility.

The plethora of personnel which one encounters here and there, is a result of the fact that often recruitment and training is lacking proper direction. In those instances where such direction is provided, it is often not comprehensive, whereas its implementation remains poor. As a result, it is not uncommon that staff members, upon completion of training, are kept waiting for six months and more before they are assigned duties, especially duties commensurate with their background and experience. One will often find staff members placed in positions where they can make little contribution to the organisation. Likewise it can be frequently observed that more than one person is assigned the same task. These circumstances may well have a negative effect on staff motivation and this occurs at almost all levels. Staff regulations should apply equally to all members and special treatment of personal favorites be avoided. Those who perform below standard should be subjected to evaluation and appropriate corrective measures. Finally and too often, a situation is created in which an employee is reduced to inactivity, for other reasons than professional ones. This gives rise to frustration, sometimes an urge to cheat and even to revenge. Personnel management is an element of primary importance to sector reform, details of which are not covered by the "green paper".

REGIONAL COOPERATION: NETWORK OF INFORMATION

A capital and knowledge intensive sector such as telecommunications can greatly benefit from regional cooperation. The existing cooperation within ECOWAS deserves to be supported and strengthened whenever possible. Very often a structure for cooperation does not exist at a national level, nor seem to function properly at the regional level. The lack of an effective cooperative structure becomes obvious when we consider the difficulties arising from the preparation of international conferences. A well functioning national and regional telecommunications committee will assume as its principal duty to centralise all documents and information on, inter alia, national and regional conferences, classify and catalogue these according to scientific discipline, thus creating a sound database.

We would recommend the creation of a computer based information exchange network, to be used by policy makers at national and regional level. Electronic distribution of information would also overcome logistical problems prevailing at subregional level in Africa, which often affect the exchange of documents through postal services and prevent direct telecommunications contact among colleagues. In addition, telecommunications policy makers are well placed to take a lead in using telecommunications in order to make their own work more efficient.

With such support facilities, the telecommunications committee would be able to offer exact briefings to the national authorities in any area of activity relevant to the telecommunications sector. In addition, African delegates to international events will also be able to better coordinate their positions. For instance, effective preparation and follow-up on the World Telecommunications Development Conference of the ITU (Buenos Aires, March 1994) would have been easier to accomplish. We are not aware of any coordinating framework which could serve for preparing this event. We sometimes see an informal "African Group" emerge and vanish at important international meetings. During the Additional Plenipotentiary Conference of the ITU in December of 1992 in Geneva, a group of this nature was formed to try and adopt common positions, on the election of the vacant post of the director of the ITU Telecommunications Development Bureau. The lack of consensus at this meeting was apparent and coordination should have taken place well in advance. A new ad hoc group was created later to prepare a draft contribution to the next ITU Plenipotentiary Conference in Kyoto (Japan) in 1994. With a telecommunications committee as suggested above, all sources of information with relevance to the Kyoto meeting can be collected for the benefit of a proper contribution.

With respect to cooperation on a technical level, it is clear that not all countries have been provided with the same new technologies, such as digital cellular radio or public services via satellite. It is desirable that a framework for cooperation is created to facilitate exchanging experiences among operating companies and other concerned parties. In order to make such cooperation effective, it is necessary to set the rules at a regional level to ensure that all regional players in the telecommunication sector assume their roles and can benefit. In practice, this would mean that a country disseminates information on important technologies which they are the first to use. This mechanism would help reduce the cost of educating personnel abroad on the same equipment and create possibilities for training in the same environment.

There are numerous complex, interesting and challenging issues related to regional and subregional legal cooperation. Each country has national legislation in place covering all sectors, including telecommunications. There are many similarities among them, particularly along the lines of francophone, anglophone and lusophone countries. While the "green paper" aims at the creation of national legal frameworks and regionally coordinated regulation, it is obvious that the first condition for change is political will. Every state cherishes its sovereignty, while telecommunication is often perceived as a strategic sector. As a result, supranational structures are not often greeted with much enthusiasm. This is one reason why African countries have to accept to be surpassed, notably by developed countries in the North. Once national interests are believed to be safeguarded, the question on what exactly could be delegated to a regional body and defining the specific limits of a state monopoly, can be addressed. Once these questions are answered, the notion of the creation of a regional legal framework with harmonised regulation can also be addressed. The two most sensitive examples in this regard are competition and privatisation.

COMPETITION AND PRIVATISATION

The first question is to define the preferred level of competition. Is full competition the better solution and why, or is limited competition more suitable and why. Likewise, when the question of privatisation is raised, the type of privatisation should be clarified, whether it is believed to solve major problems or whether it is the least unattractive option. Even a capital ownership of a public telecommunications company which puts 25 percent of shares in the hands of a foreign operator could permit him control of this company. Sector liberalisation may lead to the creation of a new private monopoly replacing the old one. Proper evaluation of reform options is preferable to *ad hoc* measures which are sometimes taken under pressure from business interests. There is a West African saying: 'Are we being rushed? Let's go slowly'. Only a well clarified political will can bring an answer to the interrelated questions of competition and privatisation.

REFORM IN CÔTE D'IVOIRE

Telecommunications sector reform is in motion and ranking high on the agenda of Côte d'Ivoire. A formal separation between the regulatory authority and the provider of public telecommunications services has been in place since 1991. However, this regulatory authority is dependent on the same government ministry which is the sole shareholder and ultimate supervisor of the operating company CI-Telcom.

In order to introduce more autonomy in this structure, an independent regulatory body is under preparation. Once the draft law to this effect, already prepared by the ministry responsible for telecommunications, is adopted, then Côte d'Ivoire will have on the one hand a regulator with wide-ranging powers and clearly separated from the shareholder, the ministry. On the other hand there will be a public telecommunications operator who may eventually face competition, even while basic telephony is to remain a monopoly service.

The liberalisation process in Côte d'Ivoire has three objectives. Firstly, the posts and telecommunications laws of 1976 will be modified and brought up to date with global trends. Secondly, regulatory and operational functions will be separated. Thirdly, competition will be allowed into the sector. In this regard, it is important to note that the development of rural telecommunications should be facilitated. We therefore suggest that regulation will create possibilities for private operators wishing to develop a rural area, spelling out obligations of all parties concerned including the national operator. A well elaborated legislative and regulatory framework is an essential condition for ensuring that a country will obtain the best solutions with the best national and foreign partners.

Meanwhile, a sound debate on the fundamental questions on liberalisation is only possible if telecommunications lawyers are available in Africa. Training of such lawyers is now more urgent than ever.

OoOoOoO

(1) Telecommunication Policies for Africa (The African Green Paper), second draft, November 1993. ITU Telecommunication Development Bureau (BDT), Geneva (Switzerland).

Telecommunications and Development in Africa
B.A. Kiplagat and M.C.M. Werner, Eds.
IOS Press

CHAPTER VII

UNIVERSAL SERVICE OBLIGATIONS FOR SOUTH AFRICA (*)

Pinky Moholi
General Manager INTOUCH Programme
Telkom
South Africa

The world eagerly awaits the developments in South Africa in light of the political social changes and the economic developmental effects. Considering the complex inequalities of the past, the time of action is now. This then is the time to lobby for effective application of telecommunications as a catalyst for economic growth especially to the traditionally disadvantaged communities of South Africa.

The overall term, "Universal Service" and its various elements are therofore at heart political and economic concepts, which are not always compatible. In order to of capture the political and economic theme, we offer to define what universal service means to South Africa:

> "A means to enable the individual, community and the nation to
> attain basic life-sustaining needs, develop self-esteem and be
> allowed to make socio-political and economic choices through
> telecommunications"

The various concepts of universal service being:

1. AVAILABILITY OF TELEPHONES - this may be defined through the lack of resources, as access to a public telephone. The value of such a service being highest where the telephone penetration is lowest e.g. rural communities. South Africa may then be divided into specific areas for service provision:

☐ *Rural Areas* - characterised by low population densities and adverse terrain. Universal Service obligations may be defined by geographic coverage i.e. a community phone per village.

☐ *Peri-Urban Informal areas* - these are the informal settlements surrounding the big cities, characterised by high population densities, unpredictable population growth and no formal infrastructure. Universal service obligations may be defined by the number of public/community phones per population.

☐ *Formal Townships* - characterised by high population densities but served with basic infrastructure. The priority is to provide public telephones, followed closely by the provision of individual telephones to anyone able to pay for one.

☐ *Urban Residential* - characterised by lower population densities with formal infrastructure, the priority is to provide a telephone point to every household willing to pay for one, with some public telephones in strategic points.

2. AFFORDABLE TELEPHONE SERVICES - determining the extent to which the low income sectors of the community are able to obtain or use telecommunications.

3. ACCESSIBILITY & USABILITY - determining the ease of access to telecommunications to all and especially the physically disabled and the illiterate users.

4. APPROPRIATENESS OF SOLUTIONS - fit for purpose technologies to satisfy the community's changing needs.

We, therefore, have to look at the environment South Africa operates from, affected primarily by:

☐ *Poverty and inequalities* - over 70% of the population subsisting on 25 % of the country's national income.

☐ *Population growth* - The present population register is 38 million and is expected to increase by 35 % by the year 2000. Telecommunications infrastructure development not only is lagging behind the population growth, but need to leapfrog in order to overcome the future demands.

❏   *Urbanization* - For economical, social and physical reasons, the normal world trends of urbanization has reacted sharply to the removal of artificial suppression of apartheid, resulting in a flood of people in the urban areas. The informal settlements have mushroomed around the cities. These are characterized by high population densities and no formal infrastructure.

❏   *Unemployment* - In 1992 the Central Statistics services reported 53 % unemployment among the black community, and 4 % of the white community.

From the above analysis, it is clear that the provision of telecommunications in South Africa is very skewed against:

❖Race with DELs/100 persons   > 75 White community
                                                      < 2 Black community

❖Rural communities with   DELs   75 % Urban
                                        Public phones 80 % Urban

resulting in the following Performance indicators:

Total DELs per 100 inhabitants   = 9.5
Public phones penetration   = 1 per 750 inhabitants

POLICY FORMULATION

The law that governs telecommunications in South Africa is outdated and amounts to a policy vacuum. Debate though, in South Africa, is active of formulation of telecommunications policy. A National Telecommunications Forum was Iaunched in November '93 to look at different aspects of policy formulation, regulatory framework and funding mechanism.

The ANC has offered a Reconstruction Program document in January 1994. The crucial priorities for the next five years being the infrastructure development viz.:

❖   Building 1 million homes
❖   Electrifying 2.5 million homes
❖   By 1996, to provide every school and clinic with a telephone, and,
❖   Make telephone services available to all.

Because universal service obligation is not driven by economic development but by political decisions, it is important to re-emphasize the role of the regulatory intervention, the national policy setting and consumer needs in ensuring the meeting of this obligation.

REGULATORY PRIORITIES

The policy priorities that need addressing are:

1. Wide spread of geographic coverage in line with ITU recommendations.
2. Rapid establishment of the infrastructure.
3. Access to affordable service.
4. Build capacity of skills to maintain the existing and developing network.
5. Set up realistic service indicators to measure performance in line with economic development.

SERVICE OBLIGATIONS

The regulator may specify universal service obligations in a broad sense, which is simple, but applies little pressure to the operators to achieve rapid solutions to universal service. The detailed long term charter regulatory direction is most desirable to South Africa in order to put more direct pressure to the operators. The first years of a democratic new South Africa, require policies that will rapidly and visibly address the needs of the disadvantaged communities. This intervention might take the form of:

i. License conditions detailing expectations with regards to universal service obligations e.g. geographic spread, contribution to development fund for new entrants and/or infrastructure development.

ii. Rate restructuring on the rental fee, and/or deferred payments on connection fees, resulting in greater participation in the public network.

iii. Free provision of access to all essential services i.e. healthline, ambulance services, security and education services etc.

PERFORMANCE INDICATORS

Performance indicators to be used to assess the current situation, to estimate how rapidly progress is made in the provision of universal service have to be quantitative.

i. Service penetration i.e. line density and geographic coverage in rural areas.

ii. Affordability.

iii. Quality of service i.e. waiting list/time for telephone line.

iv. Capital efficiency i.e. investment per line/public phone provided etc.

FUNDING MECHANISM

The sources of funds from both international and national sources should be driven by policy. International Aid will be affected by competition from other local priorities such as housing, health and education, and will be piece-meal and project based.

Nationaly funding may take the form of levies on big business with private networks, local and global VANs, new operators and other users. The re-investment of public operator's profits into infrastructure development is another source. It is important therefore, to look at what Telkom can do to generate income for the infrastructure expansion whilst keeping the leading competitive edge.

Firstly, there is no such thing as a universal product, i.e. one for which everyone is a prospective buyer. The mass market has split into ever multiplying, ever-changing mini-markets that demand a continually expanding range of options, models and customisation.

Remember the days when bath tubs were white, and telephones were black?
Today, Revlon makes 157 shades of lipstick, 41 of them pink!!

The approach of segmenting the telecommunication market in South Africa is a world trend. In this way, the service provider can address the needs of each market closer and in turn be more "in touch" with the customer; hence the branding of the Telkom community services as IN-TOUCH. The community services include:

☐ *Individual customers* - small business customers who operate a telephone bureau, a chatter box in his premises and/or sells telephone prepaid cards.

☐ *Public customers* - Coin and card phone users and community phones.

The unique aspects of the IN-TOUCH market is that there will have to be an even greater intervention by the service provider in:

1. Promotion of the services thus:

   ❖ promoting the use of telecommunications by demonstrating the advantages in both cost and time saving.
   ❖ Encouraging the development of the individual/bureau service to the prospective investor.
   ❖ Promoting selling of the phone cards by traders.

2. Education of customers

   ❖ Relevant technical and business training to the small business investor.
   ❖ Educating the customer on the use of public telephones, new card systems.

3.   After-sales support

The question every employee should ask himself is "Could I convince the person sitting next to me in a dinner party of the merit or superiority of the service my company is offering?" If you are not losing the customer you already have, if you can maintain and deepen his loyalty, that itself is a victory.

I must say, most monopolies in South Africa have a long way to go in this regard, Telkom included. but, with the threat of competition at the doorstep, the changing customer expectations, we have little choice but to change our marketing approach to meet the challenges of the new world. The growth in Telkom's customer base, the increased use of telephone services is crucial as this is offered earlier as among the means of financing the universal service obligations, which in turn, will promote economic development for the benefit of all South Africans.

OoOoOoO

REFERENCES

ITU	❖	African Telecommunication Indicators,1993
ITU	❖	The changing role of gouvernment in an era of telecommunications deregulation (brochure)
	❖	Universal Service and Innovation 3/12/93
Stan Rapp & Tom Collins	❖	Maxi-marketing, the new direction in marketing strategy
FvR Greeff, Telkom	❖	Organising for Marketing, lesson from other telecommunication countries
Acknowledgments		A. Ngcaba, ANC A. La Houd, Telkom Billy Cordani, Telkom

OoOoOoO

(*) This article is based on a presentation during "The African Horizon Conference", organized at the initiative of Telkom SA, 18 - 20 January 1994 in Pretoria, South Africa. It should be noted that universal service policies are in the process of being further developed, especially in the framework of the National Telecommunications Forum (NTF). All major sectors fo South African society are represented in NTF, who's main objective is to offer policy proposals to government and industry.

PART II

USERS NEEDS

CHAPTER VIII

AFRICA IN A GLOBAL ECONOMY - REQUIREMENTS FOR FINANCIAL TELECOMMUNICATIONS

Kees W. Hozee
Manager Carrier Relations
S.W.I.F.T.
Belgium

The financial sector is an infrastructure element of vital importance to economic development in any country. An equally crucial infrastructural role is played by telecommunications. The financial data communication and processing network S.W.I.F.T. (Society for Worldwide Interbank Financial Telecommunication), which connects over 4,000 financial institutions across the globe, relies on telecommunications to improve efficiency in the financial markets.

S.W.I.F.T. is at the centre of today's global electronic banking industry. It provides financial data communication and processing services to over 4,000 financial institutions worldwide and is in the process of introducing many of the countries of Africa onto its network.

A feasibility study recently undertaken by S.W.I.F.T. shows that by the end of 1995 at least half of the countries of Africa -- 20 out of 41 -- could be operating in the S.W.I.F.T. network and their banks benefiting from international electronic messaging services.

The difficulties involved in bringing the remaining 21 countries into the network range from low interest among the financial community to difficult or unreliable telecommunications.

GLOBAL STANDARDS FOR RELIABILITY AND SECURITY

Today the core of the world's banking community, over 4,000 financial institutions, do business with each other via the S.W.I.F.T. network. This business generates more than 2 million messages each day. These travel across public telecommunications networks to and from all the world's major financial locations.

This network has been progressively built up by S.W.I.F.T on a country-by-country basis over the past twenty years. In each case, S.W.I.F.T. has endeavoured to provide the lowest cost solution which meets the necessary standards of reliability and security.

FROM PAPER-BASED TO ELECTRONIC BANKING

S.W.I.F.T. is a co-operative owned by some 2,223 of its user banks. It was founded in 1973 by a group of 239 banks with a common goal to improve efficiency in the area of international payments. The overriding objective was to improve transaction processing efficiency by encouraging a move from paper-based procedures to electronic banking. Through a process of standardisation it has sought to overcome language and message-format barriers and to reduce processing and communication costs.

For most banks in the world, the S.W.I.F.T. way of doing business is the industry standard.

While S.W.I.F.T.'s core business is 'store and forward' message processing, it also offers a growing range of other specialised services to aid bank efficiency. For instance, it provides a netting mechanism for the forex and money markets as well as support for real-time gross settlement in large-value domestic payment systems. S.W.I.F.T. also offers bulk data transfer capabilities.

S.W.I.F.T. is also heavily involved in electronic data interchange to support paperless commercial trade. This is seen as a forerunner of much closer electronic integration between banks and their customers as part of the total value chain.

GEOGRAPHICAL EXPANSION

The S.W.I.F.T. network now covers 100 countries. These countries display many differences in terms of telecommunications infrastructure, traffic and customer base.

S.W.I.F.T. is committed to continued geographic expansion as this is part of its charter. But before deciding to invest in a S.W.I.F.T. Access Point (SAP) in a country the organisation has to be sure that it will recover its costs. Also, for a SAP (figure 1) to be installed an agreement has to be reached with the national carrier on pricing and conditions.

In addition to this, the two main challenges to implementing S.W.I.F.T. services are frequently the telecommunications infrastructure and national regulations.

AFRICA

In looking at the economies of Africa, S.W.I.F.T. has found that there are as many differences as there are countries regarding telecommunications performance, hardware supplier coverage etc.
The quality of the local telecommunications infrastructure is the basic limiting factor in the services which S.W.I.F.T. can offer.

CONNECTION OPTIONS

To bring a country into the S.W.I.F.T. network (figure 2), the preferred approach is to install a S.W.I.F.T. access point or X.25 node in the country. The member banks will generally be linked to the node through a leased line. This approach represents a substantial investment by S.W.I.F.T. which has to be balanced against the number of members who will benefit and the message revenue that will arise. Clearly in some smaller countries, it is not an economic proposition.

Today (March 1994), S.W.I.F.T. has SAPs located in Algeria, Ghana, Mauritius, Madagascar, Morocco, Ivory Coast, South Africa and Tunisia. It currently envisages establishing a SAP in Nigeria and also possibly one in Egypt (subject to regulatory approvals).

In Nigeria, for example, 26 banks belong to the African top 100 and the anticipated traffic volume justifies investment in a S.W.I.F.T. node (SAP) which should be operational in the near future. Cross-border access arrangements to these SAPs will be made for most of the other African countries.

CROSS-BORDER ACCESS AND ENCRYPTION

Though cross-border access is not the perfect solution for all countries, it will, given the lack of viable alternatives meet the strong desire of many African banks to start on the S.W.I.F.T. system immediately, and provide for considerable improvements over telex operation.

For small volume users (less than 250 messages per day or a connection time of less than 20 minutes per day), PSTN (Public Switched Telephone Network) or dialup is a good option, assuming the quality of the local lines is adequate. In such a case the connection would be to a shared PSTN group with end-to-end X.25 encryption to ensure security.

Since international leased lines to the S.W.I.F.T. access points may prove prohibitively expensive for some small volume users, the PSTN approach may be a practical alternative. A lot depends on the quality of the local infrastructure.

PUBLIC DATA NETWORKS

Another alternative currently being explored by S.W.I.F.T. is to utilise Public Data Networks (PDNs).

Currently considerable investment is being made in telecommunications in general and Public Data Networks in particular and many major cities in Africa already have nodes. This could prove a cost effective access mechanism for long-haul national and international connection to a S.W.I.F.T access point for low volume users. The potential bottleneck of this approach is access to the PDN node. If the reliability of this proves acceptable, security can be assured through S.W.I.F.T. end-to-end encryption. The security aspects of PDN, which is a critical issue, are currently being studied by S.W.I.F.T. It is highly likely that encryption will be a mandatory requirement.

SATELLITE OPTION

The use of a VSAT placed at user premises connected to a shared hub, either at S.W.I.F.T. or at a third party connected to S.W.I.F.T. is an increasingly attractive option since VSAT and user costs are falling fast. Use of VSAT allows for a dialled, on demand or a leased satellite connection.

There is currently a lack of suitable shared hub facilities and a lack of satellites with appropriate 'footprints'. It would therefore require bigger VSAT antennae.

Regulatory telecom restrictions on VSAT usage will also make its introduction on a widespread basis difficult. Current charging practices also make VSAT at present a less attractive proposition.

POTENTIAL

S.W.I.F.T. recognises that considerable future potential resides in the PDN and the VSAT technology approach but their efficiency and cost effectiveness have yet to be proven. However, with the support and cooperation of local telecom authorities, progress could be made rapidly in finding workable solutions.

In the case of PSTN (dialup), a special commitment needs to be obtained from telecom authorities to provide quality local dedicated connections between S.W.I.F.T. and the international exchange. By by-passing local exchanges, an acceptable level of operation can be expected. This approach represents a fast, low-cost start to the process.

CROSS-BORDER CONDITIONS

Where cross-border access will be used, cooperation is needed from local telecom authorities for S.W.I.F.T. to be accessed in that country. Also, banks using S.W.I.F.T. services are invoiced from Belgium and this has to conform with local fiscal regulations.

ROLE OF PTTS

S.W.I.F.T. always seeks a harmonious relationship with the national telecom authorities. The S.W.I.F.T. system represents a closed user group and offers telecom authorities new revenues from leased lines and international connection charges. These revenues are at the expense of the traditional telex traffic.

The contribution of the local telecom authorities is vital to the expansion of the S.W.I.F.T. network across Africa. S.W.I.F.T.'s experience to date has been that many national telecom authorities do not fully understand or appreciate the contribution which S.W.I.F.T. can make to their national banking systems.

Many of them appear too prepared to stick to their existing ways of operating and not look to the future. There is a general need for them to be more open and customer oriented.

In our view, the telecom authorities should be focusing their efforts on improving the basic telecom infrastructure and services. A better system will, we believe, result in considerable increases in traffic for them and, of course, revenues.

Market-oriented prices must be applied to new services such as PDN or VSAT.

A proper functioning telecommunications infrastructure is, most people agree, a basic requisite for a modern African economy. In itself it can attract more international business to a country. Already, many countries are investing in digital international exchanges which is very positive. This still leaves open the question of the quality of the local exchanges to meet the needs of the S.W.I.F.T. user.

The demand from the banks for digital message switching is overwhelming. Over 4,000 banks in 100 countries are now fully accustomed to communicating in the S.W.I.F.T. electronic format and wish to operate with their corresponding banks in places such as Africa in this way.

IMPROVING BANK COMPETITIVENESS

S.W.I.F.T.'s membership and message pricing policies are the same throughout the world irrespective of how much we have to pay for our telecommunications charges in the different locations/regions and these vary from excessive to reasonable.

The services offered by S.W.I.F.T. aim to offer user banks the means to increase their own competitiveness. This sets S.W.I.F.T. apart from other network operators such as Reuters, GEIS, IBM etc. whose owners are not their customers and whose profits do not go back to their users.

The transition from telex messaging to electronic message transfer offers banks considerable direct financial gain. The telex-based system does not provide good day-to-day visibility about financial balances. Reconciliation of accounts is often delayed resulting in potential interest loss. Also the handling charges involved in manual processing of transfers, including re-typing of telex instructions, are a further considerable cost.

Add to this the risk of errors and accounting confusion, and it is clear why most banks in the world prefer to operate using electronic transfers based around the standardised S.W.I.F.T. message formats. Most banks appreciate that, at the end of the day, using S.W.I.F.T. results in savings on fees and charges levied by international correspondents.

In countries in Africa, banks have been able to benefit from electronic message transfer by subscribing to the electronic networks of some of the large international commercial banks. This generally implies an exclusive relationship with one particular banking group which may not always be desirable for an African bank. S.W.I.F.T. offers an international network which is totally independent and offers freedom in international banking relationships.

TYPICAL S.W.I.F.T. SERVICE

S.W.I.F.T. endeavours to provide the best quality of service in terms of message delivery time. This is defined as the elapsed time between acknowledgement of receipt by S.W.I.F.T. to the sender and the availability of the same message for forwarding to the receiver.

This commitment is clearly laid out in the S.W.I.F.T. 'Responsibility and Liability' policy. Failure to meet these service levels can result in heavy penalties for us as an organisation.

The quality of the S.W.I.F.T. service is highly dependent on the services provided by the telecommunications organisations. A 'customer satisfaction' survey of S.W.I.F.T. users showed that in 1992, 32% were having problems with leased line operations. Yet the leased lines are beyond the control of S.W.I.F.T. and the telecommunications organisations that supply them do not make comparable commitments to minimum service levels for their customers.

Fortunately there are some welcome signs in this area. For instance, in Asia-Pacific where eight carriers are joining together to offer a contingency service guaranteeing 100% circuit availability, albeit with a 30% cost surcharge.

S.W.I.F.T. has been greatly encouraged by the attitude changes taking place in a number of telecom authorities in Africa who recognise the economic needs for a modern, customer-oriented telecom infrastructure and service.

Nigeria is a case in point where its positive attitude has encouraged S.W.I.F.T. to install a SAP to cater for the local demand and be a focus of cross-border business. A similar effort by telecommunications operators in other African countries could help more banks gain access to the S.W.I.F.T. network and help them meet today's international business challenges through efficient electronic communication.

NETWORK STATUS IN AFRICA:

INSTALLED SAPs (MARCH 1994):

Algeria, Ghana, Ivory Coast, Mauritius, Morocco, Madagascar, South Africa and Tunisia.

ANTICIPATED 'CUTOVERS' TO THE S.W.I.F.T NETWORK:

Kenya, Nigeria, Zimbabwe, Egypt, Cameroon, Senegal, by the end of 1994.
Ethiopia, Gabon, Gambia, Libya, Malawi, Mozambique, Swaziland, Tanzania, Zambia, during 1995.
Burkina Faso, Congo, Mauritania, Sudan, Togo, by 1996.

COMPARATIVE TELECOMMUNICATIONS COSTS

Comparative telecommunications costs in US Dollars per month			
Connection type	Leased line 2 400 bps	PSNT [3] 2 400 bps	PDN [3] 2 400 bps
USA [1]	1 300	100	600
AFRICA [2]	6 300	300	200

Note [1] :   From the West Coast to a SAP on the East Coast.
Note [2] :   From an average African country to a SAP in Europe.
           These figures will vary widely from country to country.
Note [3] :   Based on 100 messages per working day.

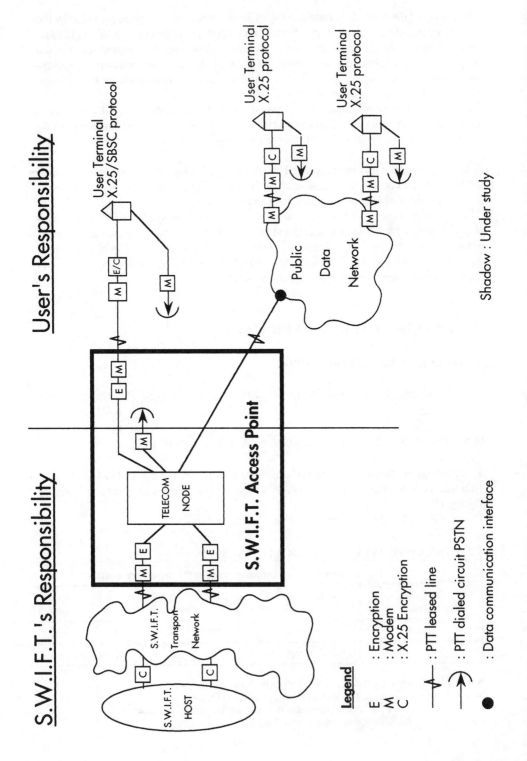

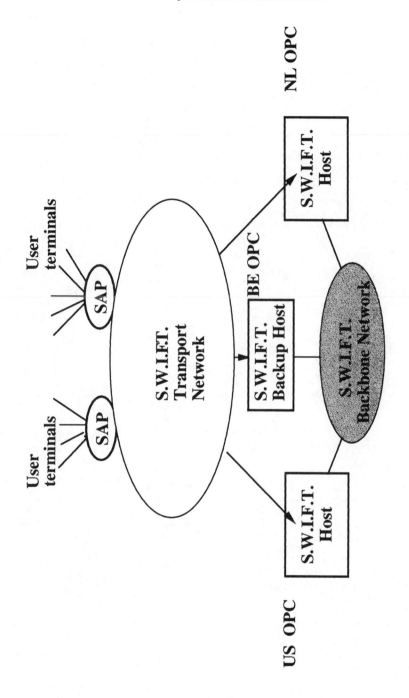

Legend

OPC : S.W.I.F.T. Operating Centre
SAP : S.W.I.F.T. Access Point

CHAPTER IX

AFRICA IN A GLOBAL ECONOMY
TAKING ADVANTAGE OF EXPERIENCES
IN EUROPE AND ASIA

Ernst Otto Weiss
Vice-Chairman of INTUG
International Telecommunications Users Group
France

INTRODUCTION

During the past decades there has been a powerful movement towards internationalisation and globalisation of the world economy. The economic relationships have continued to intensify, not only in world trade, but also in investments. Telecommunications has had a similar impact with its sustained rate of innovation. The need for massive new infrastructure investments triggered by advances in technology and demands from users for new services created critical capital demands, even for the traditional rich western TO's.

Far more critical is the situation when the expansion path is heading towards economically developing regions: the Eastern Europe markets, the Far East and - Africa. All offering almost unlimited resources of labour and material, but we must clearly differentiate between these markets, when we talk about the telecommunications environment.

While Africa is struggling for development of basic infrastructures and of demand-adequate universal services, Eastern Europe is trying to find its way out of the economic morass which is the legacy of communism, but most Far East countries have already a sustained record of technology and infrastructure advancement. However, problems are arising in all regions from high tech evolution, where finance may not in all cases be of first concern; but where cultural and religious problems, characteristics and etymology problems are still of dominating political influence.

A GLOBAL PERSPECTIVE: EASTERN EUROPE AND THE FAR EAST

Before exploring the African scenario and in order to demonstrate telecommunications development patterns, it may be appropriate to describe the advancements made in Eastern Europe and the Far East and first to provide some general remarks about what business users expect from regulators and providers of developing telecommunications services, regardless of the geographical situation of a region:

1. In relation to innovations, users have decided views on points that will need to be attended to if they are to find new offerings attractive to them. They will only migrate voluntarily to new systems if the commercial advantages are persuasive. If they do not migrate, the result will always be financial disaster to others.

2. Allowing network and services suppliers to operate in an increasingly global fashion will be the only way to create the international telecommunications environment for services which users are demanding. Further, it will help to remove the investment barriers in new technologies, and it will foster the establishment of national networks and services with harmonised interfaces.

3. Bilateral agreements on International Value Added Networks (IVANS) are starting to progress as a result of international service regulation agreements, but more scope will need to be given to much wider multi-national approaches.

4. The discrepancy between national views leads to fundamental differences in regulation and the degree of difference in open market approaches. An assessment of the true economic benefits and cost - and a common method for calculating on individual country basis - would be invaluable.

5. Universal service is an important social obligation and it offers users the potential for greater access to the marketplace and for eventual dissemination of the workforce.

6.   Small TOs may be viable to serve small countries, but it will be equally important that large, multinational TOs emerge. Such organisations will be indispensable to the development of efficient and uniform international network services.

MOVING FORWARD

This presentation can only highlight some of the broadly common terms and approaches which are adopted in carrying out telecommunications services and will deal in three parts with each of the regions separately:

CENTRAL AND EASTERN EUROPE

Central and Eastern Europe is a "region" only in a rather vague geographical sense given emphasis by relatively recent history. There is, in reality, a diversity of culture, including political culture, which is blossoming rather than fading.

The telecommunications situation is also remarkably divergent in the different countries of the region, not only in terms of infrastructure availability and service quality but also in terms of the regulatory environment. Each country is seeking its own way out of the economic morass which is the heritage of communism.

The core of this section will provide a brief overview of the common elements of the reform strategies which are now being placed in Central Europe: particularly in Poland, the Czech and Slovak Federal Republic and Hungary. It introduces some reasons why telecommunications should be given high priority in the economic reform programmes, gives an indication of the levels of investment required and presents the main elements of short term strategies for the provision of business telecommunications. This includes primarily digital overlay networks and international links, but also developments in local networking, the growth of mobile radio networks and the situation in the domain of value added and data services.

All countries in the region are facing difficulties on every imaginable level. Democratically elected parliaments and governments are working overtime to reduce the backlog of democratic legislation. The new governments of Central Europe (Poland, Hungary, Czech and Slovak Republic) are well underway in designing the legal framework for the recon- struction of their society. Progress is less impressive, but still significant in the Balkans (Bulgaria and Rumania) and the Baltic countries. In other countries, especially those of the Commonwealth of Independent States (CIS), the fundamental economic and political reforms have just been started.

Ministries responsible for telecommunications policy are adapting their legislation and regulations to reflect the requirements of a market economy. In many countries, the policymakers have recognised that the catastrophic telecommunications situation requires not only drastic actions to provide business with effective communications, but also realise that the State-owned telecommunications operator alone will not be able to face-up to the huge task which lies ahead.

But the inheritance is very heavy: Regulation of telecommunications services was mostly based on interpretation by the State administration of laws giving the State an absolute monopoly on telecommunications services, and recognising the government administration as the principal user. Therefore all countries of Central and Eastern Europe have suffered from insufficient levels of investments in the telecommunications infrastructure. Telephone services were profitable enough in the past, but most of the share of profit (e.g.80% in Czechoslovakia) was absorbed into the central State budget, rather than re-invested.

Public networks have been found to be more decrepit than anyone had expected. Many switches had not been replaced since World War II and transmission links employ mostly outdated East German or Russian analogue technology and it is generally agreed that the existing public networks require complete rebuilding, rather than modernisation and that the emphasis must be placed on meeting the needs of international business users. The short-term strategies, adopted in most countries in the past two years, particularly in the East part of the Unified Germany, reflect this.

THE REFORM STRATEGIES IN CENTRAL AND EASTERN EUROPE

All the new liberal governments are now giving telecommunications investments a very high priority. It was recognised, that business telecommunications services must become more efficient, if private investments are to be attracted and are considered a very essential prerequisite to any developing market economy

Most countries have now developed very ambitious short-term investment plans for telecommunications services, all to be realised by the end of 1993. Long-term plans have also been developed in some countries, but feasibility and financing remains uncertain.

The view that business services should subsidise residential services is promoted by the telecommunications "establishment" - mainly the TOs and the switch vendors - who have a clear interest in maximising residential penetration using profits from business services. This view of affairs is also popular with politicians who favour hidden subsidies to give prospective voters the phones they want in reasonable time.

Clearly this kind of strategy has its attractions, but it must be balanced against the need to keep business tariffs low to attract investments in industry generally. Overall, the focus is on business services and except for East-Germany and a few local experiments in Poland and Hungary, there is no committed funding for residential telecommunications development. An interesting observation, indeed!

INVESTMENT STRATEGIES

To bring Central and Eastern Europe up to the OECD average of 40 main lines per 100 population would require the installation of about 100 million main lines with the associated network equipment. Current indications suggest that somewhat less than 50 million new lines will be installed by the year 2000. If business services are given priority, about half will be for business.

Most estimates now assume that the investment required in Central and Eastern Europe is around $ 1.500 per main line, which results in a total public telecommunications investment requirement in the region of $ 75bn between now and the year 2000.

The first stage of most national plans is to install a digital overlay network, linking the major cities and provide access to international links. This approach is supported by the major lending agencies as it gives rapid access for business users to digital bandwidth and consequently the prospect of revenue and profit to pay back loans.

However it remains a key issue, facing all governments in the region, how to fund local network rebuilding and the necessary installation of local loops to residential subscribers. There are two main options:

Option 1 is to open some or all of the local networks to competitive provision in order to attract private investment in infrastructure. This poses the question of what arrangements should be made for interconnecting and billing with a variety of disparate networks. It requires a strong and visionary regulatory authority.

This model is being applied in Poland and Hungary and to a lesser exetend in the Czech and Slovak Federal Republic. As an example, five licenses for ambitious local projects have been issued by the Ministry of Post and Telecommunications of Poland. The most advanced project is that of Telekom Pila: to install and operate a totally new terrestrial infrastructure within the province of Pila and to provide telecommunications network for local access and intra-urban communications.

The license allows the offering of all telecommunications services, but obliges it to provide universal service equitable according to demand in urban and rural areas, to include the installation of payphones and to offer ISDN services in the future.

This demonstrates that impressive developments are possible - even if financing remains a problem and that the viability of many projects remains to be proven. In some ways the countries of the region have a unique opportunity to exploit technological and regulatory innovation; they are not burdened with an undepreciated investment in copper loops, nor with a political imperative to subsidise domestic rentals.

Option 2 is an alternative to the model of competition in the infrastructure and basic services level, also the traditional model of massive private and public funding of investments by the national monopoly - possible in joint ventures with Western TO partners. This model is now adopted in the Ukraine and Estonia.

Such an approach is being advocated by some Western TOs and their equipment suppliers, who would promote countries to follow (rather than leapfrog) Western-Europe developments and use their technologies. However, even if such money can be raised, its funding may strain the national economy and result in prohibitive levels of debt.

THE REGULATORY REFORMS

In most countries of the region, the most likely scenario for the next century is one of full competition in all telecommunications services. Most likely regulation will be limited to ensuring fair competition, data protection and privacy.

Until then countries will adopt their own national approaches, but the driving forces will be:

❏ Plans for early membership of the EC in some countries, which will require to carry much of EC legislation into national laws, especially separation of regulatory and operational functions and competition in telecommunications services;

❏ A political desire for replacement of the PTT monopoly - often perceived as association with previous regimes - into something more "liberal";

❏ The need to attract funds to the country for business generally and telecommunications in particular, and it is recognised that competition and partial privatisation may be conditions for loans from international financing institutions.

CONCLUSIONS FOR THE CENTRAL AND EASTERN REGION

It must be borne in mind that the reconstruction of most of Central and Eastern Europe has just begun. There are no historical precedents or models for achieveing smooth transformation from the rigid communist system to a modern democracy and a market economy. Inevitably, this is a process of "learning by doing". The failure of communism has clearly exposed the impossibility of "great designs" applying to the transformation of a whole region. The countries of the region are no longer acting as a "bloc", but are adopting their own, often highly original, approach to the challenging situation they face.

But it must be emphasised, that funding of telecommunications projects remains a fundamental problem, which is as yet unresolved. Commercial loans and loans from international institutions - which account for only 5 to 6 % of the required funding - will have to be repaid. The real sources of funding will remain internally.

The most likely sources of funding are the TO revenues, partial privatisation of TOs and locally funded operator initiatives, supported by local governments, private investors - and the business and residential users themselves.

THE FAR EAST

The decade of the 1980s has witnessed a phenomenal economic growth for the Far East, particular the ASEAN countries (Brunei, Indonesia, Malaysia, the Philippines, Singapore and Thailand). In terms of telecommunications development, provision of both basic and value-added services is on the rise with an ever greater emphasis on fulfilling user needs.

In terms of investment a greater economic and political emphasis was placed on the telecommunications sector and in the 1990s more growth is expected because of rapidly developing sophisticated networks and enhanced services. The rise in mobile systems and services, plus the growth in satellite systems will characterise the 1990s in Far East. Evidently a scheme of financing will be provided for the lesser developed countries of Indonesia, Thailand and the Philippines to produce a greater quality of basic services.

Economic cooperation has led to the development of many new relations in various sectors of economies, including the development of telecommunications systems. The new telecommunications facilities coincide with the present development of manufacturing and tourism industries. For instance Singapore Telecom entered into a joint venture development with PT Telekom of Indonesia, the Salim Group and other private concerns to develop the telecommunications infrastructure of the Batam and Bintam Islands.

Privatisation of national telecommunications authorities is currently in various stages of progress, but the need to expand beyond domestic markets is evident. Singapore and Malaysia for example are looking towards developments in Vietnam as relations with that country are soon to be normalised.

Another important trend is the increasing demand for the provision of mobile services including cellular telephony. With a rising GNP per capita, more disposable income available, multiple market operators and a better telephone line penetration; we also find that the subscriber base for cellular systems is on the rise. But a mobile telephone service with all its associated problems of standards, compatibility and tariffing will become a large regional issue as user demands are steadily growing.

COMPETITION AND COOPERATION

Both cooperative and competitive forces are beginning to shape telecommunications developments in the region. Competition in the provision of basic service infrastructures is as demanding in markets for sophisticated value-added services. In addition, the trend towards global mergers and alliances of entities with telecommunications concerns spells an even greater strategic positioning of organisations wishing to participate in the markets of the Far East.

Private organisations, both domestic and foreign, are keen to take part in this development. With the deregulation of the CPE market in most countries of the Far East, accompanied by the privatisation of most telecommunications authorities and increase in private participation, competition for lucrative contracts and various revenue sharing systems is on the rise.

But one of the most important factors in the development of telecommunications in the Far East is the diversification of user demands. Mainly the convergence of computer and telecommunications technologies has resulted in large scale data transfers requiring networks with enhanced functions to meet growing demands of multiple users.

Another major technological trend affecting this region is the merger of broadcast and telecommunications technologies. The results of this merger are just beginning to shift programming and distribution modes away from the monopoly of the national broadcaster and into the hands of private partners for the diversification of programming and operating.

THE REGULATORY EVOLUTION

The implications of regulatory changes in the Far East countries are being made within the context of the trend towards more open markets for telecommunications systems and services which follows the general economic trends. The consequence of these cumulative changes has induced a shift in policy from purely fulfillment of basic telecommunications services to the improvement of a supply of a variety of sophisticated services. The relative importance and influence of Japan, Canada, the United States and the European Community can be seen in the increasingly high competition interest for markets in the region.

The message is clear and is gaining a prominent position in discussions and debates to mandate policy changes in the Far East region. A measure of liberalisation of markets, regulatory revision and organisational restructuring will help to ensure that telecommunications can fulfill its role in supporting efficient economic and social growth. Whether deregulation and privatisation comes sooner or later, as well as the form such changes will take, will depend on the inevitable demand to create more accessible networks to meet users' needs.

Therein lies the essence of the regional development in the Far East region: because of its growth in the 1980s and plans for the 1990s, the Far East countries are increasingly able to make that development occur in the ways they deem profitable for themselves.

AFRICA

How do the above-mentioned experiences relate to the African region?

"Learning by doing" seems to be the only advisable scheme, because there is no ready-made experience available in the world which may seamlessly fit the developing African scenario.

INVESTMENT FUNDING

An obvious "chicken-and-egg" situation may irritate the neutral observer:

☐   External funding is needed in most African countries, particular in the sub-Saharan region, to provide adequate infrastructures;

☐   but foreign investors, who may provide resources for funding require adequate infrastructures to do their business.

The Eastern Europe experience demonstrates these investment problems and the hesitation of foreign investors after the first period of enthusiasm over political reforms had eased and the realities of inadequate infrastructures became obvious.

The Far East experience may point the way to overcome this problem: setting the priorities right!

Liberalisation of public telecommunications resources may be one way of solving the problem and telecom regulators around the globe are now enlightened to the need for liberalisation, as evidenced by the flood of countries reviewing and revising their telecom laws. This is certainly an encouraging step, opening the way for technological innovation and increasing pressure for efficient operation and better service offerings from TOs.

But we must be cautious and ask ourselves whether the early telecom liberalisation goes far enough in creating a regulatory and economic framework which encourages public and private investment in the administration and technology development of national, regional, but finally truly globe-serving networks.

In the next 3 years it is predicted that over 26 countries in the world - many of them in Africa - will begin to privatise their national telecoms networks, but demands for investment capital will far outstrip supply; analysts have estimated that only 10 of the 26 countries will meet the necessary criteria investors expect.

High cost, combined with fear of repatriation in unstable political environments will discourage many companies from investing in privatized telecommunications companies. Political and economic instability in many countries seeking privatisation, combined with the sometimes astronomical cost of upgrading the technologically outdated systems make these companies less attractive than initially expected.

But what to do? In view of the dependency of international trade on global working networks, business users would like to see such international governmental organisations as the ITU and the OECD become more instrumental in educating governments in Africa to provide an adequate economic climate before offering networks for privatisation.

Furthermore, business users in the world would like to advise governments in Africa to read carefully the variety of excellent material on telecommunications and world economy issues, published by business organisations such as the International Chambers of Commerce (ICC), the Telecommunications Foundation of Africa (TFA) and the International Telecommunications User Group (INTUG).

THE AFRICAN GREEN PAPER

A first attempt to structure the economic climate in Africa was recently initiated and supported during the drafting stage by the ITU/BDT.

"Telecommunication Policies for Africa - (The African Green Paper)" (1) (hereafter AGP) is a consensus document on a limited liberalisation of the telecommunications sector across Africa. The AGP reflects positions of public telecommunications companies and their respective ministries in regard to the provision of universal and public services, but does not provide enough consideration to the needs and requirements of the developing international orientation of the economies in Africa.

Seen from a global perspective and considering the experiences with the European Green Paper of 1987 it is noticed that the AGP offers a starting point for a sector reform, but here it comes to a dead stop and does not go on to suggest a comprehensive and dynamic regulatory framework for future developments. Its proposed level of liberalisation is very limited and there are no recommendations on future privatisation of present monopoly operators. We therefore conclude that it would be a major and time consuming experience for most African countries to implement such proposals.

In comparing the AGP with the equivalent European document of 1987 we find that the roots of origin are quite different - but we must also realise that the telecommunications situation in both continents is quite different.

The European document was mainly influenced by economists and politicians concerned about the threat to break antiquated monopolies in order to return to the basic rules of free market competition.

They felt that the provision of universal services had reached a level of basic saturation, but the requirements of advanced business and industry needs could not be fulfilled, thus driving the home industry into a non-competitive situation.

The AGP however was designed by African public operators in the shade of a completely different situation: the provision of basic universal services is still in many countries of the African continent at a much less advanced stage and badly needs protection of public resources to emerge, whilst the development of modern business telecommunications needs the push of entrepreneurial activities.

The document therefore concentrates on the question how to manage existing monopolies in order to foster the provision of elementary services in the region - particularly in rural areas - whilst trying to take care of the provision of adequate services for modern business expansions.

It remains to be seen which path will lead to better results in Africa, but it must be recognised that in Africa the time-clock of economic developments is running very fast.

ECONOMICS AND THE UNIVERSAL SERVICE OBLIGATION

Many controversial discussions have led to the impression that society in general and business in particular does not favour universal services. Particularly in Africa universal service is an important social obligation and to business it offers the potential for greater access to the marketplace and for eventual dissemination of the workforce.

❏ But how relevant is it that we consider the universal service obligation as a barrier to deregulation?

❏ Why should today's complicated cross-subsidisation surrounding it, create further inertia to change?

❏ Do we really understand universal service and the options for its provision all that well?

Small TOs are viable to serve small communities, but it may be equally important at the other end of the scale for a large, regional TO to emerge. Such organisations may be crucial to the development of one-stop, uniform availability of universal and international network services. Who otherwise will have the resources to ensure quick adoption of new technologies and shoulder the increasing investment cost of telecom R&D.

But as the OECD suggests (2), there is no institutional reason why universal service should not be required of and delivered by either public monopolies or regulated private competitive operators.

The question as to which mode is preferable from the universal service perspective rests upon the economics of delivery. Clearly "bypass" and "cream-skimming" are potential threats to the revenue base of existing monopoly suppliers of telephone service, which make it increasingly difficult for them to provide service to domestic subscribers in difficult areas, e.g. rural areas, at an acceptable price.

But the fact that the debate on this question is far from being resolved in the developed world points to the needs for flexible and non-dogmatic policy responses to this question in the developing world, particularly in Africa.

THE ROLE OF GOVERNMENTS

But not only for Universal Services, the responsibility for the development of the entire telecommunications market in Africa lies firmly on the shoulders of national governments, and they have a challenging task ahead to balance the emphasis of their various roles as:

❖   regulators,
❖   suppliers and competitors,
❖   major users,
❖   influencers of technical standards
❖   guardians of public interest.

We believe that rapid and effective development of any telecommunications framework can only be successfully achieved by the continued reduction of legislated constraints in favour of a more open telecommunications marketplace. This demands a light framework of regulation - concentrating on market principles and not on rules. It will be important for governments to continue divesting their TO interests and to be sensitive to market distortions created by their position as major users.

Governments must strive individually and in concert to achieve an open telecommunications market and withdraw to the role of market watchdog by:

❖   ensuring fair trade conduct and practises,
❖   overseeing the market, prices and performance.

Anything more technologically oriented is unrealistic - since the framework will not keep pace with scientific advance and user demand - and will therefore be restrictive.

The immediate future for Africa promises to become an era of exciting business and technological innovation and change. Governments need to move quickly to ensure that they create regulatory frameworks which are more service, performance and business-conduct related. We believe that the combination of technology pull and user thrust will solve the technology problems if a climate which encourages innovation is in place.

The area of concern is surely the regulatory scene. Only few of the African governments are showing signs of awareness of the user need for efficient tools to help them to be competitive in the global market which they now face

Those who have started to evolve regulatory regimes in which competition can flourish are probably less than 10% of the African continent. The telecommunications scene is bedevilled by restrictions rooted in the monopoly era in which it was born, and not enough is happening to offer hope of a change in the foreseeable future.

We are looking hopefully at the OECD to give a lead in promoting telecommunications liberalisation in Africa, because we believe they are in a better position to do this than other international fora which study the subject. OECD members are primarily interested in economic and financial aspects of world affairs, whereas most others represented by their TOs have different and more limited objectives.

I would like to focus my closing remarks on the theme that users more perhaps than any other interest group, have a vital stake in a solid telecommunications future, which concludes that they should therefore have a more participative role in the regional and national policy development in Africa than is the case at present. What they have to say and the experience they have to contribute, is essentially the kind of advice that can save huge sums in misconceived investments and expensive wrong turnings - and of which is no lack of examples in the developed world.

You will not expect me to go into details about user needs in relation to the wide range of existing new telecommunications projects now within the scope of becoming commercially important. Satellite developments, ISDN/ATM etc. are of course major examples.

But in relation to these advances, users in Africa will have decided views on points that will need to be attended to if they are to find new offerings attractive to them. They will only migrate voluntarily to new systems if the commercial advantages are persuasive. If they do not migrate, the result is financial disaster to others.

I end therefore by expressing the conviction that it is very much in the overall interest of the African community and the efficient use of limited resources, that users should be closely involved in the early stages of telecommunications changes as there is a major economic pay-off which will result from close and trusting liaison between users, manufacturers and providers.

If this sounds like a platitude, I can only say that we still have to go a long way before the brave world of tomorrow becomes reality in Africa.

OoOoOoO

ANNEX 1

ABBREVIATIONS:

INTUG: International Telecommunications User Group, a global organisation of Business Users.

OECD: Organisation for Economic Cooperation and Development. An intergovernmental organisation of the 24 economically most developed countries in the world.

TO: Telecommunications operator (organisation). The provider of public telecommunications services.

PTT: Postal, Telegraph and Telephone Organisation. Old expression for a governmental department (or ministry) acting as its nation's common carrier.

GNP: Gross National Product.

ITU/BDT: International Telecommunications Union / Telecommunication Development Bureau.

ICCP: Information, Computer and Communications Policy. A directorate of the OECD (see above).

REFERENCES:

(1)   Telecommunication Policies for Africa (The African Green Paper), second draft, November 1993. ITU Telecommunication Development Bureau (BDT), Geneva (Switzerland).

(2)   see OECD/ICCP Publication Nr.23: Universal Service and Rate Restructions in Telecommunications, OECD Publications, Paris 1991.

Telecommunications and Development in Africa
B.A. Kiplagat and M.C.M. Werner, Eds.
IOS Press

CHAPTER X

AFRICA AND SCIENCE
THE AVAILABILITY OF COMPUTER
COMMUNICATIONS

Dr Paul Godard
Gondwana Project
Université Catholique de Louvain
Belgium

This document addresses the actual status of electronic networking in Africa from the user's viewpoint. How useful are African telecommunications networks for computer communications purposes? What are their limitations? Which services do they offer? What is the cost of an individual connection to a global computer communications network such as the Internet?

The article introduces the major electronic networking services, the most common management protocols in use, and the hardware & software requirements for an individual user. Then, the current status of electronic networking in Africa is described in detail in terms of networks and services availability; comparative maps and table are given. Following a number of recent recommendations on electronic networking in Africa, the Gondwana Project addresses aspects of awareness and points to specific action for strengthening computer communications capabilities in the continent. Gondwana intends to establish an electronic communications network and information system for research, development and training on sustainable systems of land and resource use in Africa. As such, the project poses a model approach for other academic and social or business sectors as well.

It should be stressed that whereever a basic telephone line is available, many international computer networks can be accessed. As a result, even a rural telephone extension can bring vast productivity gains to local institutes using computer resources. Electronic networking poses a strong case for promoting rural telecommunications.

INTRODUCTION

Africa's poverty and development problems could be eliminated to a large extent through the use of science and technology as the driving force of economic and social progress.

On the other hand, a good scientific or technological base needs a sound information system to sustain it. Although the African region is known for its weakness in information systems, the communication infrastructure can vary from very good to abysmal from one country to another.

Recent developments in microcomputer technology and communication software make it possible to envisage the provision of a comprehensive and efficient system for the dissemination of information. Personal computers are increasingly available throughout Africa and, with the addition of a modem and communication software, they can transfer data over poor quality telephone lines at minimal cost. These recent telecommunication technologies can benefit Africa immediately, even before global networking is systematically available. Low cost solutions are already available to give individuals or individual institutes access to electronic networking.

Electronic networking refers to any form of information exchange between computers through various method of interconnection. Advantages are multiple: speed and inexpensiveness (many messages in one phone call), reliability of transmission (error-correcting modems) and ease of set-up and use. A wealth of scientific and business information is easily accessible from sources all around the globe. Why being left unaware, why re-invent the wheel, when there are so many people at reach who are prepared to share information?

ELECTRONIC NETWORKING SERVICES

There are two basic categories of services:

❖ Computer mediated communications which allow people to exchange messages;

❖ resource sharing, offering access to computing resources such as files and databases.

Either type of service may be interactive, when messages are delivered and read immediately, or batch when messages are received after a certain delay.

COMPUTER MEDIATED COMMUNICATION (CMC)

CMC services may be primarily either one-to-one (electronic mail), one-to-many (mailing lists and bulletin boards systems - BBS) or many-to-many (conferencing systems) users.

ELECTRONIC MAIL (E-MAIL) - Electronic Mail (e-mail) allows an individual user to 'mail' a message to another user who is registered on the same or an interconnected network. The message is delivered to the addressee's registered electronic mailbox either immediately when all the links are interactive, or after a delay that might reach up to several days depending on the location. Speed of delivery and reading the message is also influenced simply by the frequency at which the addressee checks his mailbox. Communication can be immediate in the best case, when the user is continuously connected to his network and receives an automatic in-box notification on the screen of his personal computer. The reliability of e-mail varies considerably, especially when network boundaries are crossed. Nevertheless, most network systems usually inform the sender when messages can not reach their destination.

MAILING OR DISTRIBUTION LIST - Networks that support e-mail by individuals to individuals often support mailing lists. These distribution lists involve groups of people who want to hold extended discussions on the same subject. When a user is interested in a subject discussed on a specific mailing list, he has to subscribe by sending an e-mail message. Then, when his subscription becomes active, he receives all the messages posted to the list. These mailing lists are normally supported by the same software as used for e-mail. Although this feature may look like a real advantage, it can easily become a problem of information flooding, since incoming mail requires some minimal processing by the user.

COMPUTER CONFERENCING SYSTEM - The computer conferencing system differs from the mailing list in scale, number of people within a conference group and in the total amount of conference groups. One copy of a message is kept per central host computer, rather than one message per individual user as is the case with e-mail. Automatic separation of messages into categories are usually supported by the system software. A true conferencing system can thus display lists of categories and lists of subjects of messages for each category in which the user may select and avoid messages by subject, sender or any logical combination of these and other attributes.

BULLETIN BOARD SYSTEM (BBS) - As BBS's usually have a limited number of topics, unsophisticated user interfaces and are rudimentary single machine conferencing systems run by a system operator (sysop), they should be considerated as intermediate systems — positioned in between electronic mailing lists and conferencing systems — where users post messages as if on a physical pegboard and with no real idea of who will read them or reply to them.

RESOURCE SHARING

Remote Login - Most interactive networks support remote login which is the use of a network to access a remote computer as if one were logged in on it from a local terminal.

FILE TRANSFER - The ability to get a file from a remote host and put it back is called file transfer. The major file transfer services is file transfer protocol (FTP) — and its associate anonymous FTP for which the user does not need an access password. The most generally usable file transfer format is plain text in 7 bit USASCII. Although this service is faster and easier to use in interactive mode, it is also supported on top of e-mail as batch file transfer for some public databases or information libraries.

REMOTE DEVICE ACCESS - This service provides a way to use devices — such as printers, tape drives, CD-ROM's — on other systems as if they were connected the the local system.

MANAGEMENT PROTOCOLS - To keep complexity manageable, protocols are designed in layers, building up from those near the hardware to those near the user. The International Standardisation Organisation reference model Open Systems Interconnection (ISO-OSI) has 7 basic layers : physical, data link, network, transport, session, presentation and application.

The major protocol stacks used in wide area networks (WAN) are the Internet (which layers organisation is close to the ISO-OSI model), the coloured book (UK & Australia), MAP/TOP of General Motors and Boeing, XNS of Xerox, DNA of Digital, NCA of Apollo and SNA of IBM. These protocols are designed with the assumption of dedicated links between nodes and networks. Among the protocols adapted to intermittent connections (dialup protocols), the most popular ones are UUCP Fido, Kermit and Xmodem.

INTERNET PROTOCOL STACK (TCP/IP) - The TCP/IP (Transmission Control Protocol/Internet Protocol) protocol stack is certainly the most implemented of the vendor independent protocol stacks, and is available on computers ranging in size from supercomputers to personal computers. The main goal of this protocol is to develop a robust communication architecture to accommodate multiple types of communications services over a wide variety of networks. TCP/IP is being widely used in South Africa, as well as over some leased lines between Africa and USA/Europe.

UUCP - The UUCP (Unix-to-Unix Copy Protocol) is most commonly used in dialup connections. The g protocol is used over the public switched telephone network (PSTN), the f protocol on the public data network (PDN), and the t protocol over TCP/IP. UUCP is largely used in Africa, especially today since it allows an easy transition to TCP/IP which is definitely the full Internet solution.

FIDO PROTOCOL - Fido protocol is similar to UUCP and uses adaptations of Xmodem and Zmodem. This protocol is largely used in FidoNet (see below). Fido is currently the number one protocol in use in Africa. It is very reliable, even over poor quality telephone lines.

KERMIT - Kermit is an error correcting file transfer protocol usually used in manually dialed connections.

XMODEM - X-, Z- and Y-modem have the same function as Kermit. These protocols including Kermit are widely used within communication packages.

HARDWARE & SOFTWARE REQUIREMENTS

Assuming that personal computers are widely available in Africa, the only additional hardware needed is a high speed modem with compression and error detection/correction protocols (V32bis & MNP10) to overcome the often poor quality of telephone lines. The price range of such modems is between US$ 300 and 500. The choice of the communication software depends on the transport protocol in use at the central computer node which one will be using (Fido, UUCP or TCP/IP). Most of these programmes are 'freeware' and can be downloaded free of charge from different electronic servers worldwide.

AFRICAN ELECTRONIC NETWORKING TODAY

The community of electronic networkers in Africa is still relatively quite small. However, through the pioneering efforts of dedicated individuals — establishing projects, improving and disseminating communication software, training of users — electronic networking has gained a foothold in Africa. The success in some pilot projects — such as ESANET, CGNET, Healthnet, NGONET, PADISnet and RIONET — have raised much interest in African networking and a large-scale project such as RINAF.

NON-PROFIT NETWORKS

Most of the non-profit network purposes are either to link research and academic institutions, healthcare centres and hospitals, or NGO's.

ARSONET - ARSONET is a CIDA (Canada) professional development project to link the Africa Regional Standards Authorities in Addis Ababa (Ethiopia), Nairobi (Kenya) and Cairo (Egypt) with Fido networking technology.

CGNET - CGNET (Consultative Group Network) is a conferencing system specialised in agriculture. It was founded in 1985 by the CGIAR and is funded by FAO, World Bank and UNDP. The CGNET host computer is a Digital VMS system located in Palo Alto (California, USA) that connects 130 hosts (agricultural research centres, CGIAR centres, UNDP offices) in 70 countries. The network consists of about 200 mailboxes for 10,000 people , all connected via the public data network when available or the public telephone network. In Africa, CGNET links the following organisations: IITA in Benin, IMI in Burkina Faso, CIP in Burundi, CIP, IITA and NCRE in Cameroon, ACDI, Ford, FAO, ICARDA, IDRC, IRRI and PLAN in Egypt, ILCA in Ethiopia, FG in Ghana, IBSRAM, IIRSDA, RTI and WARDA in Ivory Coast, CIP, Ford, ICIPE, ICRAF, ICR, IDRC, ILCA, ILRAD, PLAN, Rockefeller, TSBF, UNEP and USAID in Kenya, CIAT in Malawi, IER and ILCA in Mali, IAV, IIMI and USAID in Morocco, ICRISAT and IMMI in Niger, IITA and ILCA in Nigeria, CIP in Rwanda, IDRC, IIEN and Winrock in Senegal, CIAT and PLAN in Tanzania, FG and IFDC in Togo, CIP in Tunisia, CIAT and ICRAF in Uganda, and CIMMYT, FAO, ICRISAT, ILCA and PLAN in Zimbabwe. Based on Dialcom system software, CGNET offers other services such as databases, airline reservation and interfaces to fax and telex. Useage fees may be high for an individual: basic rate at $ 50 (host) + $ 5 / mailbox (user) per month and connection rate at $ 6-9 per hour + $ 0.095 per 1 Kb ($ 0.2 / page).

EARN - EARN (European Academic Research Network) was formed in 1983, following the example of BITNET in the USA. Its charter states that it is a network for Europe, the Middle East and Africa. EARN hosts exist in Algeria, Egypt, Ivory Coast, Morocco and Tunisia. EARN links more than 600 hosts, corresponding to about 30,000 users. Many links to national and international networks exist. EARN is funded by each participating country but is not charging the individual user.

ESANET - ESANET (East and Southern African Network) is a pilot project to link researchers at universities in Kenya, Tanzania, Uganda, Zambia and Zimbabwe with each other and with researchers worldwide, by installing e-mail facilities at their computer centres. Zambia, Kenya and Zimbabwe can connect directly to the GreenNet gateway, while Uganda and Tanzania can only connect via Nairobi because direct dialling facilities outside the (Preferential Trade Agreement (PTA) area are not available. Zambia has begun to experiment with direct dialling to London, wheras the other nodes are expected to begin testing connectivity later. Based at the University of Nairobi, Institute of Computer Science, ESANET is partially funded by the Nirv Centre (Web) in Toronto, Canada, and co-ordinated with the NGOnet project to allow NGO's who do not own host computers to use central computer resources. The system is based on Fido software running on PC's with high speed modem and dedicated phone line.

FIDONET - FidoNet started in 1984 in the USA. FidoNet is a very popular and informal network chain owned and operated by end-users and hobbyists who pass messages from node to node. It is a point-to-point and store-and-forward e-mail system, using dialup telephone links. It has over 13000 public nodes with 1,000,000 users of e-news and 100,000 users of e-mail. Usually a Fido node is a single caller BBS, however, some can admit up to 20 users simultaneously. The average is 200 active users per Fido BBS. FidoNet offers gateways to Internet and UUCP. Initially based on Xmodem for MS-DOS personal computers, Fido technology now uses Zmodem and is also available for Unix and Apple machines. It is the most affordable e-mail and e-news system used by many NGO's throughout Africa. Henk Wolsink is the Zone 5 co-ordinator <electronic address: hwolsink@catpe.alt.za>.

GHASTINET - The GHASTINET (GHAna national Scientific and Technical Information NETwork) offers e-mail service in Ghana to FOE, GAPVOD and other institutions. It uses the Fido protocol and connects twice a week to the GNFido gateway in London. Charges are $ 0.57 per page.

GREENNET - GreenNet started in the UK in 1986 as a conferencing system. It is now member of the APC. The network regroups academic and research institutions as well as NGOs in 31 African countries. In most of the French speaking countries — Burkina Faso, Cameroon, Congo, Ivory Coast, Egypt, Mali, Madagascar, Mauritius, Niger, Nigeria, Senegal, Seychelles and Togo — it links to ORSTOM nodes (see RIOnet). Elsewhere, it links to Devel Workshop in Angola, AlgeriaNet, University of Botswana, the African Development Bank in Ivory Coast, PADIS in Ethiopia, the

ACHRDS in the Gambia, FOE-Ghana and GhastiNet, the ELCI and the University of Nairobi in Kenya, the National University of Lesotho, the University of Malawi, the CIUEM in Mozambique, the University of Namibia, ENDA-Dakar in Senegal, the University of Swaziland, the COSTECH in Tanzania, IRSIT in Tunisia, the Makerere University in Uganda, the UNZA in Zambia and the University of Zimbabwe and MANGO in Zimbabwe. GreenNet uses Fido protocols and links to WorkNet/SANGOnet and UniNet in South Africa.

HEALTHNET - HealthNet is operated by a Boston based NGO called Satellife which was initiated as a project of the IPPNW. The network was initially addressed to exchange health and medical information within the universities participating in the ESANET project and via Memorial University in Newfoundland in Canada. HealthNet uses store-and-forward micro satellite called HealthSat to pick up and deliver e-mail messages and electronic publications for the health sector. The ground station uses technology that is affordable and appropriate for African conditions. Ground stations are operational in Cameroon, Congo, Ghana, Kenya, Mozambique, Tanzania, Uganda, Zambia and Zimbabwe, and licensed in Botswana, the Gambia, Malawi, Mali and Sudan. Although traffic is currently limited to health related issues, it will be up to the individual participating institutions in Africa to obtain clearance from the authorities for a wider interpretation of the health mandate. As far as the funders of HealthNet are concerned, this could encompass a much broader range of environmental and social issues.

INTERNET - The Internet facilitates sharing of resources at participating organisations — such as government agencies, educational and research institutions, and private corporations — and collaboration among researchers, as well as to provide a testbed for new developments in networking. The Internet was formed in 1983 when ARPANET, its oldest constituent network (1969), was split to form MILNET. The Internet is very large and a very important facility for researchers around the world to exchange information. Initially located in the USA, it now extends into Canada, Europe, the Philippines, Korea, Japan and recently also into Africa, notably in Algeria, Egypt, South Africa and Tunisia. At present the Kenya Computer Institute is working towards establishing an Internet node in Kenya, which will be shared and co-sponsored by the University of Nairobi, the Kenyatta University, RICOSIX-WHO and some other NGO's. Internet links are also under preparation at EMI in Morocco and at the Addis Ababa University for the CLEO networking project. The Internet is composed of many academic networks all running the TCP/IP protocol stack. Its major component networks are PRNET, AMPRNET — used by short-wave radio amateurs (hams), DDN, NRI, CSNET and NSFNET. The Internet has recently developed links to other major networks, such as ACSnet, ARISTOTE, EUnet, JUNET and NORDUnet. Although the Internet is often referred to as "The Worldwide Network," the following well-known international networks are not part of it: BITNET, EARN, HEPnet, JANET, NetNorth, USENET and UUCP.

NGONET - NGONet is funded by IDRC. The MANGO project is a Fido bulletin board service located in Harare, Zimbabwe. It is operated by a collective of NGO's : the Africa Information Afrique (a regional news agency), EMBISA (a religious development group), SARDC, EDICESA and SAPES. MANGO connects three times daily with the Web gateway in Toronto. In addition it connects three times a day to WorkNet in Johannesburg.

PADISNET - Initially, from its inception in 1980, PADIS (Pan African Development Information System) network proceeded by non-electronic means. It was established to assist African countries in strengthening their national capacities for collection, storage, and utilisation of data on development, and to promote information exchange in Africa. Later in 1990, PADISNET was formed as the largest of all African networking projects and supported by IDRC. The PADISNET links 37 African countries into a network of development planning centres which exchange databases and information. PADIS is based at the UN-ECA in Addis Ababa (Ethiopia) which operates the PADISNET node connecting on demand to London, South Africa and the USA. NGONet and PADISNET share resources in the support of nodes in Dakar (Senegal), Accra (Ghana) and Dar Es Salaam (Tanzania). PADISNET is based on Fido technology. The local PADISNET BBS at Addis Ababa is organised around four items: PADIS news, ECA news, data communications issues and general news on information systems and technologies. PADIS also offers public on-line access to the databases maintained on its HP3000 host computer. As a FidoNet node, PADISNET participates in several on-line conferences such as consultation conferences of experts in the African region (EXPZONE5) and the global NGONET conference.

RINAF - The RINAF (Regional INformatics Network for AFrica) project was conceived by the Intergovernmental Informatics Program (IIP) of UNESCO, financed by a grant of the Italian Government and a contribution from Korea. The project started in 1992. The technical administration is under the auspices of the CNUCE institute of the Italian CNR in Pisa and a supervisory African Committee has been created. The project is meant to bring basic Internet services to several African countries by establishing 5 regional nodes — CERIST in Algeria, NCST and Moi University in Kenya, NCTM and Obafemi Awolowo University in Nigeria, CNDST in Senegal and University of Zambia, several nodes in Cameroon, ENSTINET/FRCU in Egypt, Ethiopia, Gabon, Université de Conakry in Guinea, Ivory Coast, Mozambique, Manzini University in Swaziland, Tanzania and Tunisia. RINAF relies on the co-operation with other existing initiatives operating in the African continent such as the IDRC and the RIOnet projects.
The protocols used are heterogeneous — Fido, UUCP and TCP/IP — because of the variety of interconnections. The Project Co-ordinator is Stefano Trumpy <electronic address: trumpy@vm.cnuce.cnr.it>.

RIONET - The Paris based research organisation ORSTOM has involved itself — through the RIOnet (Réseau Inter-tropical d'Ordinateurs) — in the setting up of a network that presently interlinks many laboratories in tropical countries including 11 French speaking African countries — ESI, ARTS, CECI and UERD (Ouagadougou and Bobo-Dioulasso) in Burkina Faso, ENSP and OCCGE (Yaoundé) in Cameroon, Brazzaville in Congo, Abidjan in Ivory Coast, CIMAD (Antananarivo) in Madagascar, ISFRA, INRSP, CERPOD, WHO/OCP, IER and PNVA (Bamako) in Mali, Maurice in Mauritius, SEAG, AGHRYMET and CERMES (Niamey) in Niger, ISRA, ENSUT, CSE, CORAF, the Université CAD de Dakar, the Ministère Sénégalais de la Modernisation and the Agence Panafricaine de l'Information (Dakar) in Senegal, Victoria in Seychelles and Lomé in Togo.

International organisations are also partner of the RIO project: CIRAD, OSS, GRET, EPH, UNITAR and the World Bank.

RIOnet is member of the Internet. The central node is located in Montpellier, France. The RIOnet links LAN of Unix workstations and of Apple and IBM-compatible PC LANs, set up in ORSTOM and at its partners' laboratories. It uses available telecommunication links (dialup lines or X.25 network) and TCP/IP or UUCP (g & f) protocols. The RIOnet counts 800 accounts (1,000 users) among researchers, engineers, technicians and administrative staff. Users pay a cost depending on the quantity of data; local access is free and intercontinental transit ranges from $ 0.35-070 / 1 Kb. The Project Manager is Pascal Renaud <electronic address: renaud@orstom.fr>.

SANGONET: See WorkNet.

UNINET - UniNet is the South African university TCP/IP network connected to the Internet via a 64 Kb digital leased line from Rhodes University to IMCnet-Atlantic in Washington DC.

It connects via UUCP dialup lines to other universities in Botswana, Lesotho, Malawi, Mozambique, Namibia and Zimbabwe.

The major networking centres in South Africa — Cape Town, Grahamstown, Durban, Johannesburg and Pretoria — are connected via 64 Kb digital leased lines. Other links range from 9.6 Kb analogue lines to 64 Kb digital ones. UniNet is only funded by participating universities. Telkom, the South African national telecommunations company, imposes regulations that no other traffic but university flows through UniNet. The contact person at FRD is Vic Shaw <electronic mail address: vicshaw@frd.ac.za>.

USENET - UseNet is a e-news system which counts more than 2,500 newsgroups. African netters can access UseNet except for the picture/sound files.

WEDNET - WEDNET supports research on women and natural resource management. WEDNET is based at ELCI in Nairobi. Its aim is to link researchers in

Senegal, Ghana, Burkino Faso, Nigeria, Sudan, Kenya, Zimbabwe, Zambia and Canada via electronic communications and conventional networking.

WORKNET - The network has been established in 1990 and now has about 150 users on a multi-user BBS programme called MajorBBS. WorkNet operates as the national electronic network host for NGO's in South Africa, called SANGOnet. Users include the labour movement, human rights groups, the alternate press, documentation centres, service organisations and church groups. SANGOnet runs a proprietary store-and-forward network format. By converting the internal format to Fido format, users can send and receive messages to other systems and obtain conference postings. It is connected to the GreenNet Fido gateway via London (high speed modem and X.25 leased line), as well as to NGOnet via MANGO in Harare (daily dialup). The contact person is Paul Nash <electronic address: paul@frcs.alt.za>.

COMMERCIAL NETWORKS

AFRIMAIL - Afrimail is a bilingual (French and Arabic) e-mail system. It has been initiated at CNI and CIRIA of Tunisia in collaboration with UBC and IDRC both in Canada. Based on the Ean software, the system uses PDN, leased lines, PSTN as well as and telex lines. It links other countries through UUCP from the host tuniscni.

AT&T-MAIL AND MCI-MAIL - Both AT&T-Mail and MCI-MAil are worldwide commercial e-mail networks, originally based in the USA.

COMPUSERVE - Compuserve is a worldwide commercial network based in USA. Based on the EasyPlex software for e-mail, Compuserve also offers BBS databases mostly related to computer technology and business services. The service interfaces with fax and telex. The basic useage rate is $ 9 per month including 60 e-mail messages. Internet e-mail access is available. Although this seems a good bargain for some African countries, there are very few local dialup access numbers in Africa (Kenya, South Africa and Djibouti).

SAPONET - SAPONET is the X.25 PDN carrier run by Telkom in South Africa; it is not an e-mail network. It provides routing to Beltel, a teletext system similar to the French Minitel.

TELKOM400 - Telkom (South Africa) is running a commercial X.400 e-mail service, known as Telkom400. A e-mail passerelle exists from and to Internet.

TWICS - TWICS (Two Way Information Communications System) is a commercial e-mail service serving Japan primarily but providing a link to South Africa. It also links to the USA.

TYMNET - Tymnet, a commercial Public Data Network, has links in South Africa.

Tab. 1 : Networks (*Internet, BITNET, UUCP, FidoNet & Satellife*) and cost of 1-minute phone call from Africa to London, New York & Paris and from Bruxelles to Africa for all African countries.

Country	ISO	Inter	Bit	UUCP	Fido	Sat	London	NY	Paris	Bxls->
Algeria	DZ		B		F					1.54
Angola	AO				F					3.33
Benin	BJ						6.99	7.47	4.82	3.33
Botswana	BW			U	F		3.85	4.67	3.85	3.33
Burkina Faso	BF			U	F					3.33
Burundi	BI									3.23
Cameroon	CM	I		U						3.33
Cape Verde	CV						5.16	3.64	3.64	3.33
Central Africa	CF									3.33
Chad	TD						6.02	7.73	4.02	3.33
Comoros	KM						5.62	9.04	4.02	3.33
Congo	CG			U		S	7.03	11.24	5.62	3.33
Djibouti	DJ					S				3.33
Egypt	EG	I	B	U	F					3.23

Country	ISO	Inter	Bit	UUCP	Fido	Sat	London	NY	Paris	Bxls->
Equatorial Guinea	GQ									3.33
Eritrea	ER				F		3.82	4.86	3.57	3.33
Ethiopia	ET				F					3.33
Gabon	GA						6.63	15.66	5.22	3.33
Gambia	GM			U	F					3.33
Ghana	GH			U	F	S	2.73	2.73	3.54	3.33
Guinea Bissau	GW									3.33
Guinea	GN									3.33
Ivory Coast	CI		B	U	F					3.33
Kenya	KE			U	F	S	3.38	3.38	3.38	3.33
Lesotho	LS			U	F		3.56	4.27	3.56	3.33
Liberia	LR									3.33
Libya	LY									1.54
Madagascar	MG			U	F		7.57	9.45	4.06	3.33
Malawi	MW				F		1.98	1.98	1.98	3.33
Mali	ML			U	F					3.33
Mauritania	MR						2.91	3.88	1.58	3.33
Mauritius	MU			U	F		1.73	2.02	1.73	3.33
Morocco	MA		B		F		1.80	3.60	1.80	1.54
Mozambique	MZ			U	F	S				3.33
Namibia	NA			U			1.92	2.26	2.25	3.33

Country	ISO	Inter	Bit	UUCP	Fido	Sat	London	NY	Paris	Bxls->
Niger	NE			U						3.33
Nigeria	NG			U	F					3.33
Reunion	RE	I		U						3.33
Rwanda	RW				F					3.23
Senegal	SN			U	F		4.42	5.89	3.53	3.33
Seychelles	SC			U			4.49	4.49	4.49	3.33
Sierra Leone	SL									3.33
Somalia	SO									3.33
South Africa	ZA	I		U	F					3.33
Sudan	SD					S	5.78	4.22	5.78	3.33
Swaziland	SZ			U	F		2.27	3.00	3.00	3.33
Tanzania	TZ				F	S	8.00	8.00	8.00	3.33
Togo	TG			U						3.33
Tunisia	TN	I	B	U	F		1.17	2.43	1.02	1.54
Uganda	UG				F	S	4.78	3.98	4.78	3.33
Western Sahara	EH									3.33
Zaire	ZR									3.23
Zambia	ZM				F	S				3.33
Zimbabwe	ZW			U	F	S	1.92	1.92	1.92	3.33

Telecommunications and Development in Africa
B.A. Kiplagat and M.C.M. Werner, Ed.
IOS Press

CHAPTER XI

DEFINING PERFORMANCE INDICATORS FOR A POST-APARTHEID SOUTH AFRICA

Dr Tim Kelly
Head of Operations Analysis, Strategic Planning Unit
International Telecommunication UnitITU[1]
Switzerland

BACKGROUND

The elections to be held in South Africa on 27 April 1994 will herald a new era of political, social and economic development. Nowhere is this more true than in the field of telecommunications. The ITU has been working with the Centre for Telecommunications and Information Policy Development (CDITP) to define a series of telecommunication performance indicators which could be used by the proposed new regulatory authority in South Africa. Once defined, this system of indicators could equally be used in other African nations with a similar distribution of population between urban areas, rural villages and semi-formal settlement patterns.

[1] The opinions expressed in this article are those of the author and do not necessarily reflect the opinions of the ITU or its Member countries.

THE TELECOMMUNICATIONS ENVIRONMENT IN SOUTH AFRICA

The current state of telecommunications development in South Africa reflects a history of institutionalised inequality between white and non-white races. This can be seen quite clearly for instance in the levels of teledensity in the different cities of South Africa which vary much more widely than for other countries at a similar level of development (see Figure 1). Teledensity (number of telephone mainlines per 100 population) varies between almost 30 lines per 100 people in Pietermaritzburg and less than 10 in Vereeniging. Differences in the provision of telephone service to different ethnic groups are even greater. A study in the Durban area showed that the average level of penetration in white households is 77 per cent, but in non-white households it ranges from 2 per cent of households in the informal sector (predominantly black) to 28 per cent among coloured races. For this reason, the development of performance indicators for South Africa is all the more important in order to document the current state of the network and to assist the operator and the new regulatory authority in developing priorities for resource allocation.

South Africa currently has just over 3.5 million telephone main lines giving a teledensity of just under 9 telephone main lines per 100 population. South Africa thus has the highest level of telephone penetration in Sub-Saharan Africa, but this is still less than a fifth of that achieved in the OECD countries. South Africa's recent network growth has been modest, at around 7 per cent per year over the last decade. But, as Figure 2 shows, the pattern of growth has been patchy ranging from 10 per cent per year in the mid-1980s to just 3 per cent in the last two years. This is barely sufficient to keep up with the rate of population growth. The growth patterns of Telkom, the national operator, is typical of state-owned PTTs in that its investment patterns has fluctuated according to government macro-economic policy. Even though Telkom was "corporatised" in October 1991 and formally separated from both the postal services sector and the Ministry of Transport, Posts and Telecommunications, nevertheless it remains constrained in its ability to raise capital on the open market.

One of the underlying reasons behind Telkom's poor growth performance in recent years is its high level of indebtedness. Since the mid-1980s, Telkom has been disbursing as much as a quarter of its revenue just to keep up with interest payments in its debt which was inherited from the government at the time of the corporatisation (Figure 3). In consequence, Telkom's investment has declined markedly in the second half of the decade, particularly when the statistics are corrected for South Africa's double digit inflation rate during this period (Figure 4).

Figure 1: Mainlines and teledensity variations in South African cities, 1991

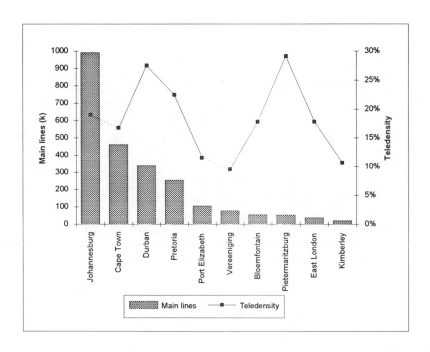

Figure 2: Mainlines and growth rates in South Africa, 1980-92

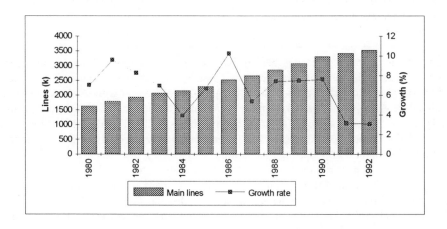

Figure 3: Telkom's failure to maintain investment growth ...

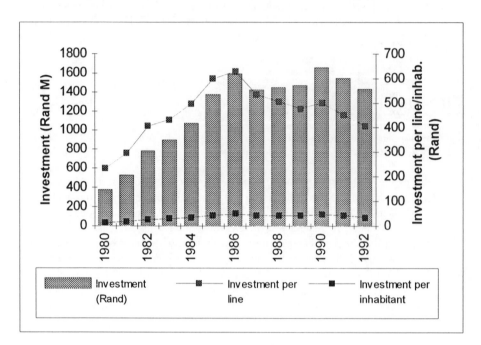

Figure 4:   ... and the reasons why

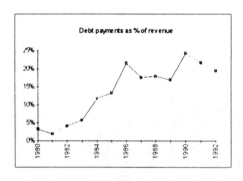
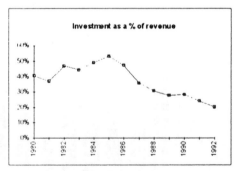

Telkom's investment has declined not only in absolute terms but also as a percentage
of revenue. Telkom currently reinvests less than 20 per cent of its annual revenues.
Experience from elsewhere in the world indicates that this figure should be at least 40
per cent and preferably higher if high levels of network growth are to be sustained.
One of the reasons for Telkom's debt is that it has historically been paying a relatively
high amount per new line added. Since 1981, almost two million new lines have been
installed at an average cost of US$4000 each. This compares with a world market

price of around US$1500. There are a number of reasons for this, for instance the geography and sparse population density of South Africa raises the cost of each line. But the main reason is probably the effect of sanctions and the fact that Telkom has historically purchased lines from the local subsidiaries of Alcatel and Siemens on long-term cost-plus contracts with relatively little competition. The contracts prohibited the suppliers from selling abroad thus losing possible opportunities for economies of scale.

OPTIONS FOR CHANGE

Discussion at the seminar focused on various options for raising the level and efficiency of Telkom's investment. Ideas included:

·   Rebalancing Telkom's tariffs. At present, the local call charge is relatively low (18.9 SA cents for five minutes) and the local call zone covers a relatively wide area (50 km radius). Given that more than 88 percent of calls are local, this tariff structure is failing to generate sufficient revenues for investment. On the other hand, long distance call charges are relatively high with a cost ratio between local and long distance calls of more than 50. This acts as a deterrence to inter-urban traffic. Tariff rebalancing, which could include a rise in fixed charges relative to usage charges, should help Telkom to become more self-sufficient in its ability to raise money. At present, the average revenue per line is around $200 compared with a global average of US$750 per line per year.

Figure 5: The effect of sanctions
New lines added and their investment cost, 1980-92

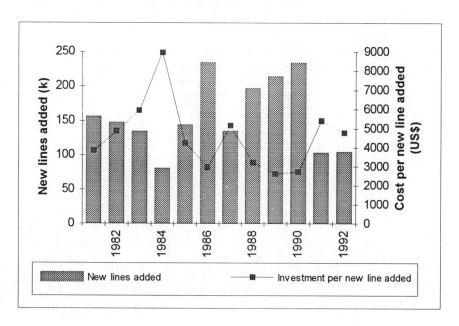

- Allowing Telkom to raise money on the international capital market. Even though it may seem strange to advocate more borrowing when Telkom is already highly in debt, it is clear that Telkom's past investment record has been highly erratic because of the constraints on government macro-economic policy. Allowing Telkom to raise money in commercial capital markets would give it more control over its investment programme and should provide lower interest rates.

- Moving to open tender for equipment procurement. The long-term supply contracts with Alcatel and Siemens begin to expire next year. This, together with the dropping of sanctions, should allow Telkom to move towards competitive tendering for switches and other equipment.

- Liberalising the market for equipment rental. At present, supply of the first telephone handset is a Telkom monopoly. Liberalising this market will allow Telkom to concentrate its investment on switching and transmission equipment rather than terminal equipment.

- Permitting greater private sector investment. In 1993, for the first time, the principle of private sector supply of telecommunication services was permitted when two digital cellular radio (GSM) licences were awarded. Telkom was awarded one licence in partnership with Vodafone (UK). The other went to MTN (Mobile Telephone Networks) a consortium of Cable & Wireless, Rembrandt (a local tobacco company), M-Net (the commercial TV provider) and two local enterprises. In the longer term, it may be necessary to introduce some private sector participation in Telkom, if only as a way to reduce the level of indebtedness.

- Government commitment to telecommunications. The new government after April 1994 is likely to be dominated by the ANC. They have already demonstrated their commitment to establishing an independent regulatory authority. They may also be willing to take on some of Telkom's debt. However, the priorities of the new government are likely to be in areas such as education, health, water, affirmative action, housing and enterprise development rather than in telecommunications per se.

OPENING UP SOUTH AFRICA

One of the most interesting features of the telecommunications environment in South Africa is the way in which political change is mirrored in South Africa's pattern of international traffic. Over the last four years, the volume of international traffic has increased from 80 billion minutes in 1989 to 220 million minutes in 1992, a growth rate of 40 per cent per year, with a particular leap in 1991 when the ANC was unbanned. This has been achieved despite the fact that South Africa has some of the highest international tariffs in the world and is some 40 per cent more expensive for international calls than the average for OECD countries. The changes in South Africa's political situation are also apparent in its calling pattern. Traffic with other

African countries, particularly with South African countries such as Namibia, Botswana, Zimbabwe, Swaziland and Lesotho, is increasing at a much faster rate that calls with South Africa's traditional trading partners in the USA, the UK and the rest of Europe. This suggests that South Africa is becoming more "African".

Figure 6: Political change reflected in telecommunication change
Growth in international traffic from South Africa, 1989-92

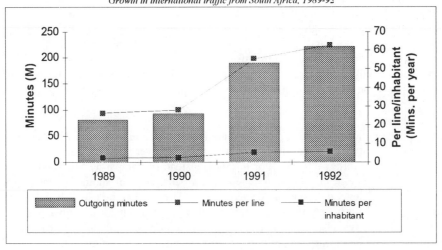

Figure 7: The "Africanisation" of South Africa
South Africa's major calling partners, 1992, and change 1991-92

Country	Minutes (1992; M)	% of total	Growth (1991-92)
UK	37.2	19.6%	+11%
Namibia	27.3	14.5%	+75%
Zimbabwe	18.5	9.7%	+6.6%
USA	17.6	9.3%	+6.6%
Botswana	12.3	6.5%	+13.6%
Africa	92.6	48.9%	+26.5%
Americas	21.2	11.2%	+8.2%
Asia	3.6	1.9%	-0.5%
Europe	66.3	35.0%	+7.4%
Oceania	5.4	2.8%	-0.2%

NEXT STEPS

As noted above, one of the main purposes of this project is to define telecommunication performance indicators for a post-apartheid South Africa. The study will be undertaken by two post-graduate researchers from the Centre for Development of Information and Telecommunication Policy (CDITP) under the guidance of the ANC and the ITU. Most of the indicators discussed in this paper were designed to compare South Africa's position with the rest of the world. However, it will be necessary to develop new indicators to measure the level of service provision within South Africa, in particular between regions and between ethnic groups.

Some of the indicators used can be the same (e.g. telephone lines per inhabitant, per household, quality of service, tariff comparisons). However, it would be necessary to adapt them to a different scale. For instance, instead of comparing South Africa to the global average, it may be preferable to set a benchmark of another country of a similar size and level of economic development. Also, it may be useful to take one region of the country, say the Cape, and use this as a benchmark for other regions. Tariff indicators will also need to be adapted to take into account ability to pay. One useful way of measuring the waiting list, for instance, is to estimate the "cost of ownership" of a telephone based on rental charges plus local calls plus a few long distance calls . As a rule of thumb, cost of ownership should not exceed 3 per cent of a families monthly income. On this basis, it is possible to calculate what percentage of the population of a city or a region could afford to own a telephone if the service were available. This form of calculation invariably produces an estimate of the waiting list which is much higher than that officially recorded.

As well as these existing indicators, other new indicators will need to be developed . For instance, it would be useful to measure levels of customer satisfaction measures for different regions or different ethnic groups. This would complement the more technical measures of service quality such as faults per line or fault repair times. The project will also need to consider what obligations for reporting indicators should be imposed on Telkom and other competitive operators such as MTN. It is likely that the ITU will continue to play a supervisory role in developing these indicators and in implementing a monitoring system once a new regulatory authority is in place. Once a system is in place, the next objective will be to extend the system of performance monitoring and benchmarking to other countries of the region which might request assistance. In this sense, the role of regional co-operation will be crucial and here again ITU is working with regional organisations such as the Pan-African Telecommunication Union (PATU) to strengthen regional co-operation.

OoOoOoO

Telecommunications and Development in Africa
B.A. Kiplagat and M.C.M. Werner, Ed.
IOS Press

CHAPTER XII

LOOKING AT UNSATISFIED DEMAND IN SOUTH AFRICA

Hans A. Koning, Steven W. Blees
University of Amsterdam, Faculty of Communication Science

The telecommunication market of South Africa is in a period of great change and national politics are the market maker in the country.

South Africa has a relatively large network with a high degree of digitization. Therefore South Africa is likely to be affected by international market forces in the telecommunications sector. The business community started pushing for privatization and liberalization, eager for better-quality service and more competition. Telecommunications has traditionally been a sector where state intervention took place. Government control became looser in October 1991 when the national operator Telkom SA was established as an autonomous limited company owned by the State.

South Africa is not just any country. The way in which the telecommunication infrastructure was allocated to the population is skewed due to the policy of apartheid (*Morris & Stavrou). Allocation is primarily focused on the white urban areas followed by white rural areas, whereas black urban areas received only little priority and black rural areas were almost totally neglected (see figure 1).

figure 1

	white population	black population
urban	++	+/-
rural	+	-

+ = allocation of service
- = no allocation of service

The question arises whether South Africa with its unequal allocation of access to the telecommunication network has, from an economical point of view, satisfied the demand for main lines of the country as a whole.

Government policy has three main objectives: (1) improve the telephone penetration in the townships and rural areas; (2) service provision at the lowest possible price level; (3) increase service level to internationally competitive standard. Now is the time for South Africa to make a choice which direction is to be followed concerning access to telecommunications. The country could choose a social policy, focused on:

❖ universal geographical availability
❖ non-discriminatory access
❖ reasonable costs or affordability

On the other hand South Africa could follow international trends and decide to deregulate and liberate the market, aimed at:

❖ reduction of state-involvement
❖ higher efficiency

This paper analyzes whether South Africa, taken as a whole, has been able to fullfill the demand for main lines over the last 10 years, as could be expected on the basis of Gross Domestic Product (GDP) per capita. The answer will enable to make recommendations on telecommunications policy.

The demand analysis will be applied to a further four countries in Africa: Botswana, Gambia, Ghana and Mauritius. The results are shown in graph A and the values are presented in the appendix.

THE CENTRAL QUESTION OF THIS PAPER IS:

Has South Africa as a whole, from 1981 - 1990, been able to satisfy the demand for main lines which could be expected on the GDP per capita?

In order to determine whether a country has been able to satisfy the demand for main lines, 30 countries were analyzed on the relation between GDP per capita and teledensity. The equilibrium was determined with logarithmic correlations. This equilibrium can be interpreted as the regression curve where GDP-based demand has been met. Countries with a higher penetration of main lines than could be expected on the economic indicator are countries with satisfied demand. Countries lying under this regression curve are countries with unsatisfied demand. A few notes must be made on applying this method to South Africa.

It seems clear that South Africa has not been able to satisfy demand in certain regions or socio-economic groups. Our analysis cannot determine whether a country positioned on the regression-line has been able to satisfy all individual demands. The analysis looks at the teledensity of a country as a whole and does not consider distribution of main lines inside a country, due to lack of reliable data.

According to the *principal of relative constancy*, media spending tends to be constant over time relative to the general performance of the economy. The distribution in which people of different countries purchase media-products does not necessarily have to be equal. Therefore, consumers in different countries with the same economic position may show a different demand for main lines.

UNIVERSAL SERVICE

Universal service is commonly understood as access to the telecommunication network. Universal service can be regarded as an economic good, in which demand and supply control the market. This means that when the teledensity of a country is lower than could be expected on the bases of the GDP per capita, the country has unsatisifed demand, and therefore has not been able to reach universal service. When a country has a higher teledensity than could be expected on the basis of its GDP per capita, the country demonstrates satisfied demand and has reached universal service. In our analysis the penetration rates in combination with the GDP per capita determines whether a country has been able to fulfill existing demand.

This analysis does not follow the concept of universal service by which access to the telecommunication infrastructure is a basic right of all citizens, to be guaranteed by government.

THE RELATION ECONOMY - TELEDENSITY

Economists have always been interested in potential growth factors in the economy. The various elements of economic growth interact in a complex manner. The role of telecommunications in this process is difficult to determine.

In the early sixties A.G.W. Jipp published a study describing the correlation between teledensity and GDP per capita. Jipp was stimulated by the study of N.J.H. Jones (1958) that found that teledensities and GDP per capita of several industrialized countries were grouped around a trend line. In addition to Jones, Jipp showed that over a longer period this trend line is virtually unchanged. The various countries move upwards on the trend line.

The Hudson study, carried out as an ITU-OECD project, showed that the usage of telephones increased with economic development. The more sophisticated and differentiated a national economy, the more telecommunications are being used. A second finding was that the use of telecommunications varies with the economic structure of the society. When the dominant economic activity is the services sector, the use of telecommunications is greater than when the economy is dominated by manufacturing industries. Manufacturing and industries require more telecommunications than the primary sector, agriculture. However, this does not imply that a telephone is more important for a bank manager than a farmer. Analyses focused on multiplier effects of the telephone at different stages in the development of the telecommunication system demonstrate that the contribution of telecommunications, and more specifically of the telephone, to economic development is particularly high in low income countries (Hardy & Hudson, 1981).

Talking about the relation between economics and teledensity, there has always been debate on the question of which triggers which. An example of the Spanish teledensity developments show that economic development leads teledensity. In the 'Global Telecommunications Review', a Paribas Capital Markets Report, an analysis was made of a number of regions in Spain. A graph incorporating all data showed that even the less prosperous regions applied to the relation between teledensity and GDP per capita, implying that line penetration is a function of GDP growth. If it were the other way, then line penetration for expanding countries would be higher than suggested by the size of the economy. However, network growth will almost always increase economic activity. There is a reinforcing effect from the teledensity on the economy, but the dominant factor is GDP per capita.

Some reports give the relation between teledensity and the GDP per Capita in a logarithmic curve. This shows quite clearly the marginal utility of the GDP per capita on teledensity growth; an increase in GDP per capita in a low income country causes a greater teledensity growth than in a high income country (Hardy & Hudson, 1981).

METHODOLOGY

The main source for our data on South Africa was the 'Yearbook of Common Carrier Telecommunication Statistics' of the ITU (19th edition, 1981-1990). The data offered in this publication are provided by national PTO's.

The standard for comparing (un)satisfied demand used in this study consists of an analysis of 30 very different countries. Together those countries are the reference point on which South Africa was analyzed. The data file contained teledensities and GDP per capita for 30 countries for the period 1972 through 1990. GDP's were converted to the 1990 US$ rate. A regression analysis was made for the 30 countries. The analysis showed a high correlation (r = .89929) between teledensity and GDP per capita (*).

After the relation between the two variables was established, the next step was to see what the position of compared to the thirty countries was. For each year a regression analysis was applied to South Africa, which resulted in annual different values for both variables (*see appendix). The formula used for this procedure is presented in figure 2. In this formula log exp td is the logarithmic value of the teledensity that could be expected on that years GDP per Capita.

figure 2

$$\textit{log exp td} = (\textit{log gdp/cap} \times a) + b$$

THE SOUTH AFRICAN CASE

The teledensity expected on the basis of GDP per capita is calculated for each year with the regression analysis, and South Africa was positioned in respect to the equilibrium. The results of these calculations are presented in figure 3.

The results show that for the period 1981 through 1990 demand has been unsatisfied almost throughout: the actual teledensity is lower than what could be expected on the basis of GDP per capita. The only exception is 1989 which shows satisfied demand. The expected teledensity for 1989 is 8.61 percent, which is less than the actual 9.04 percent.

Figure 3 : Satisfied and Unsatisfied Demand for Lines - South Africa

year:	GDP/CAP 1990 US$:	actual teledensity:	expected teledensity	satisfied	unsatisfied
				demand	
1981	1033	7.17	8.69		
1982	1156	7.89	8.74		
1983	1224	8.00	8.52		
1984	1428	8.16	9.05		
1985	1565	8.42	9.57		
1986	1743	8.36	9.27		
1987	1984	8.53	9.25		
1988	2270	8.71	9.00		
1989	2589	9.04	861		
1990	2738	9.74	10.85		

If there is a satisfied demand, as in 1989, GDP per capita is lower than what is suggested by the actual teledensity. A plausible explanation for the 1989 GDP decline are international sanctions against South Africa.

ECONOMIC OUTLOOK AND TELEDENSITY

The continued weak economic performance in the industrialized world and the likelyhood of political uncertainty following April 94's elections makes it unlikely that the South African economy will recover quickly: for 1994 a GDP growth of only 1.2 per cent is forecast (see figure 4).

Figure 4

	1994	1995	1996	1997	1998
GDP %:	1.2%	2.3%	2.7%	2.9%	3.1%
unemployment:	48%	48.9%	49.8%	50.7%	51.7%

source: The WEFA group, jan 1994.

Two periods must be distinghuished for an economic outlook, 1990-1993 and 1994-1998. The first period shows negative growth. GDP per capita will decline and South Africa will move towards satisfied demand in teledensity. The second period is expected to show a positive GDP per capita growth, causing unsatisfied demand again.

Prior research confirms that despite negative economic growth a country with unsatisfied demand can demonstrate teledensity growth. This phenomenon will continue untill an equilibrium of satisfied demand is reached. Based on the relation between economic development and penetration of main lines, it is to be expected that South Africa will show unsatisfied demand untill the year 2000.

SOUTH AFRICA'S CHOICE

South African policy makers have to make basic choices about how the telecommunication network should be developed. There are two possible options: either commercialization of telecommunications or telecommunications as a public service.

The global trend in telecommunication is commercialization and privatazation of PTO's. This commercialization has serious implications for the provision and distribution of telecommunication services. The 'business network' will be extended with various services, with tariffs based on competition. The provision of services in the rural areas is most likely not profitable and will therefore be limited to the most needed connections. This scenario will not contribute directly to welfare in black rural areas and on an increase of teledensity. A change in the unsatisfied demand situation is not to be expected in this scenario.

The other option for policy makers is continuation of telecommunications as a public service. In such a scenario the operator would be obliged to provide services and universal service is regarded as a basic right. Providing telecommunications services in less profitable areas is not attractive from an economic point of view. The expansion of a network to reach the entire population requires a social approach, in which a 'natural' monopoly seems essential.

THE SOCIAL SCENARIO

A social scenario means that universal service is to be interpreted as a basic human right. Reasonable costs and a non-discrimatory access will be the main drivers for the telecommunication policy. These two conditions are expressed in the definitions of universal service by the OECD (1991). This policy can only be maintained when telecommunication is regarded as a public service.

UNIVERSAL SERVICE IN THIS SCENARIO IS DEFINED AS FOLLOWS:

Acces to telecommunication services as a basic right of all citizens, essential for full membership of the community, and as a basic right to freedom and expression.

The rural areas, peri-urban informal areas and formal townships are the major geographical areas to benefit from this approach. A socially driven policy means high investments and, at least for several years, low revenues. It will however be a considerable contribution to the unlocking of important parts of the country and the black communities.

THE COMMERCIAL SCENARIO

The emphasis in the commercial scenario is on economic reform, whereby government influence is reduced. Commercialization is instrumental to activate this scenario.

Universal service in this approach is defined as follows:

Universal service is an economic good, to be consumed in a market like any other.

This definition emphasizes the economic point of view, controlled by the principal of demand and supply. Telecommunications policy will focus on commercialization, which will result in more and advanced services at competitive rates for the business community. Revenues per line are likely to increase and the total network will be profitable. The commercialized network is only benificial to the urban (business) community. The three geographical areas described in the social scenario will find it hard to benefit from this high-quality network.

CONCLUSION

The social and the economic scenario are interrelated. Socially driven policy tends to conflict with the market orientated objective in the economically driven policy. Increasing telephone penetration and affordability will be more realistic if connecting charges and tariffs are kept at low levels. This is conflicting with the economic driven policy, where profitability is the main objective.

As long as one dominant operator, Telkom SA, is serving the South African market at large, cross-subsidisation will be a necessity to make the social policy manageable. However, the policy aimed at commercialization could imply total privatization of Telkom SA. This policy focused on business users will keep the current unequal allocation of telecommunication resources, and the situation of unsatisfied demand, intact.

After analyzing South Africa, the (un)satisfied demand situation in Botswana, Gambia, Ghana, Mauritius was examined as well. They all show unsatisfied demand for the entire decade. In order to meet the demand those countries should have on the bases of their GDP per capita, a social policy seems more appropriate than a policy of commercialization.

OoOoOoO

REFERENCES:

Hardy, A, & Hudson H, 1981, The role of the telephone in economic development: an empirical analysis. Keewatin Communications, Washington.

Horwitz, R.B., The politics of telecommunications reform in South Africa, In: Telecommunications Policy, may/june 1992, pp. 291-306.

ITU, 1983, Yearbook of Common Carrier Telecommunication Statistics, 10th edition, Geneve, 1983.

ITU, 1992, Yearbook of Common Carries Telecommunication Statistics, 19th edition, Geneve 1992.

Morris, M.L, & S.E. Stavrou, Telecommunication needs and provision to underdeveloped black areas in South Africa, In: Telecommunications policy, sept/oct 1993, pp. 529-539.

Parker, E.B., 1981, Economic and social benefits of the REA Telephone Loan Program, Equatorial Communications, Stanford.

WEFA Group, 1994, World economic outlook, volume 1.

OoOoOoO

Satisfied and unsatisfied demand for the year 1990

Country	GDPCAP	actual TD	Expected TD	Unsatisfied demand	Satisfied demand
Morocco	1028	1.60	5.14	3.54	
Turkey	1733	12.21	7.65		4.56
Chili	-	6.16	-	-	-
Mexico	-	6.05	-	-	-
Brazil	-	6.26	-	-	-
Colombia	1112	7.32	5.45		1.87
Poland	-	8.62	-	-	-
Argentina	-	11.0	-	-	-
Portugal	4325	22.6	15.37		7.23
S-Korea	5526	31.52	18.53		12.99
Ireland	-	27.9	-	-	-
Spain	-	32.35	-	-	-
Italy	18288	38.69	46.17	7.48	
Singapore	12181	38.67	33.86		4.81
Belgium	19715	40.11	48.89	8.78	
Greece	4958	38.51	17.06		21.45
Hong Kong	-	42.76	-	-	-
Austria	19408	41.8	48.31	6.51	
Australia	14975	-	39.64	-	-
Netherlands	18041	46.24	45.69		0.55
France	21690	48.18	52.58	4.4	
Norway	23706	50.17	56.27	6.1	
Finland	25136	53.42	58.84	5.42	
Cananda	19165	57.0	47.85		9.15
USA	21700	50.88	52.6	1.72	
Denmark	21804	56.57	52.8		3.77
Switzerland	34052	58.0	74.18	16.18	
Sweden	23495	68.08	55.9		12.18
S-Africa	2738	9.74	10.85	1.11	

source: ITU, Yearbook of Common Carrier Telecommunication Statistics, 19th edition, Geneve 1992.

```
                    ┌─────────────────────────┐
                    │   Actual : Unsatisfied   │
                    │         demand           │
                    └─────────────────────────┘
                               ⇓
                    ┌─────────────────────────┐
                    │   The social, political, │
                    │   economical dichotomy   │
                    │     of South Africa      │
                    └─────────────────────────┘
              ⇓                              ⇓
   ┌──────────────────────┐      ┌──────────────────────┐
   │     Scenario I        │      │     Scenario 2        │
   └──────────────────────┘      └──────────────────────┘
              ⇓                              ⇓
   ┌──────────────────────┐      ┌──────────────────────┐
   │  Telecom as Public :  │      │  Telecom as economic  │
   │       Service         │      │         good          │
   └──────────────────────┘      └──────────────────────┘
              ⇓                              ⇓
   ┌──────────────────────┐      ┌──────────────────────┐
   │    Beneficial for :   │      │    Beneficial for :   │
   │  * formal townships   │      │  *urban residential in│
   │  * peri-urban informal│      │  casu white, business │
   │    * rural areas      │      │                       │
   └──────────────────────┘      └──────────────────────┘
              ⇓                              ⇓
   ┌──────────────────────┐      ┌──────────────────────┐
   │  Starting a more equal│      │    Continuation of    │
   │ distribution of resources│   │maldistribution of resources│
   └──────────────────────┘      └──────────────────────┘
              ⇓                              ⇓
   ┌──────────────────────┐      ┌──────────────────────┐
   │ Move towards satisfied│      │    Continuation of    │
   │        demand         │      │  unsatisfied demand   │
   └──────────────────────┘      └──────────────────────┘
```

PART III

REGIONAL COOPERATION

Telecommunications and Development in Africa
B.A. Kiplagat and M.C.M. Werner, Eds.
IOS Press

CHAPTER XIII

COMBINING FORCES : THE CHALLENGE TO CONNECT AFRICA

George R. Langworth
Director of Marketing, AFRICA
FLAG Limited
USA

International connectivity of the African continent is about to receive a major boost: the African countries are currently receiving proposals of historic proportions from three different parties for the construction and exploitation of a new submarine fiberoptic cable along its entire coastline. Below are the details of one of these proposals.

ITEM 1

Since last June, 1993, Alcatel has been working with several West African telephone companies in an attempt to develop a West Coast Africa cable. They attempt to justify the long term financing of this cable through a rationale based on connecting large populations in Africa, from a combination of vendor subsidies, French ECA (Export Credit Agency) loans, and borrowings from donor agencies by the West African countries.

ITEM 2

In October, 1993, FLAG first proposed a strategy for Africa, whereby it would work with African countries to develop connnectivity strategies to FLAG's primary configuration — a 31,000 km link from the UK to Japan — in order to provide new permanent facilities between African telephone companies and many new correspondents. FLAG will then work with these African countries to grow these new bearer circuit facilities into E-1s.

This evolution from new connectivity to wideband will be powered by targeting business communications. Shared telecom facilities will be implemented to generate new demand for business communications to joint ventures in developing African countries. These shared facilities will be strategically located beside cable terminal station sites for coastal Africa and also beside the international gateways of interconnecting facilities to inland Africa. In response to this new wideband communications demand, FLAG proposes to finance and build an international submarine cable around Africa. Each African telephone company will be able to purchase circuits as required. To support the African telephone companies as full-partners in this international connectivity, non-recourse financing facilities funded by commercial lenders can be organized around a series of landing points of the FLAG Africa cable, covering the African telephone companies' purchase cost of international half-circuits, implementation costs of the shared telecom facilities, as well as the cost of implementing all elements of connectivity required between the new international circuits and the shared telecom facilities.

ITEM 3

In February, 1994, AT&T has proposed that a cable be built around the entire continent utilizing vendor subsidies, and pre-construction capacity commitments from large multinational corporations — to be combined with long-term financing from the African countries' donor agencies.

NOW THAT THERE ARE THREE SEPARATE INITIATIVES TO CONSTRUCT A FIBEROPTIC SUPERHIGHWAY AROUND THE CONTINENT OF AFRICA, IT IS MORE LIKELY TO HAPPEN THAN EVER.

However, in order to follow through on this momentum we have to sort out the three parties' strengths, as well as the optimum strategy which will utilize these strengths in a complementary mix.

Alcatel can be regarded primarily as a submarine cable and telecommunications equipment manufacturer. While AT&T also fulfills this functionality admirably, they pursue two other lines of business which effect their potential roles in a Pan African fiberoptic cable. AT&T plays an important role as one of the biggest international carriers of the world, sporting more connectivity options, corresponding with more carriers, and participating in more international tariffs than another other carrier. AT&T has also organized a global consortium of international carriers which have banded together under the banner, WORLDREACH. The strategic objective of WORLDREACH is to utilize a worldwide network management facility to capture a considerable market share of the large multinational businesses who depend on global networks in the conduct of their business.

FLAG is a telecommunications system integrator which invests in international submarine fiberoptic cables, procuring their manufacture and implementation by vendors such as Alcatel Submarcom and AT&T Submarine Systems according to FLAG's specification. Then FLAG will operate this cable and sell its circuits to telephone companies as they need them. In order to market its circuits successfully, FLAG must also present its customer with the lowest cost solution possible in any terminal pair connection. Also FLAG employs this strategy to connect investor nations with developing nations as a means of ensuring demand for its circuits from telephone companies.

FLAG (Fiberoptic Link Across the Globe) is a two phase strategy, whereby an East-West submarine cable will be followed by an African cable descending down the east and west coasts of the continent. FLAG is a subsidiary of Nynex Network Corporation of New York (USA), with three partners Gulf Associates (USA), Dallah Albaraka Group (Saudi Arabia) and Marubeni Coropration (Japan). FLAG is entirely privately financed.

WHAT ARE THE COMPARATIVE ADVANTAGES?

AT&T has stated that if African countries purchased the cable directly from them then they would pay only for the cable. However, if Africa countries purchase the cable from FLAG, they would have to not only pay for the cable, but fund FLAG's profit as well. AT&T has characterized FLAG's role as a 'carriers' carrier' and as the addition of a new layer of administration and overhead to the construction and administration of the international cable facility. A similar argument can be made for the Alcatel position. FLAG's proposal is simply to offer a low cost solution for the telephone company's international connectivity requirements.

The comparison of the financial consequences of these three alternatives assumes two sets of constant factors:

a)   the construction of a 60,000 circuit 2 fiber pair cable around the continent of Africa, with a length arbitrarily assigned at 20,000 kilometers and an equivalent cost per km to that of FLAG — the resultant cost assumption amounts to $800 million for a pan-African cable which conforms to these length and cost assumptions. This construction cost assumption equates to a construction cost of $13,333 per bearer circuit, end-to-end. The annual stand-by operations and maintenance charge per circuit of this cable has been assumed to be 4.5% of its capital cost per annum.

b)   A revenue projection based on the actual 10 year traffic forecasts of a series of African telephone companies, extended at similar forecasted growth rates for the design life of a fiberoptic cable. The tariff assumption is based on the African telephone companies' share of the average tariff to call into Africa from Europe, $1.80 [50%] of $3.60 per minute.

The analysis evaluates the approach of both AT&T and Alcatel which entail the assumption of 100% of the construction cost ($800 million) and maintenance charges (4.5% per year) upon the cable's completion and ready for service date. The analysis assumes that long term financing (25 years) will be available for this obligation at an interest rate of 8%. Each additional point of interest assumed will add $8 million per year to the carrying charges of the obligation. During the term, debt service has been assumed to consist of interest only until the last four years of the term when the principal would be retired in four equal payments.

INVESTMENT IN ACCORDANCE WITH TRAFFIC FORECAST

The analysis compares the AT&T and Alcatel long term financing to the approach of FLAG, which entails the assumption of circuit costs from only those circuits purchased by the African telephone companies in accordance with their traffic forecast requirements. The market pricing model for the end-to-end bearer circuits of this pan-African cable which was used in this analysis is that of FLAG's primary configuration.

While the assumptions utilized in this analysis were certainly arbitrary, nevertheless they were based on clearly analogous precedents to the elements of this project. Furthermore, the true facility of this analysis is in the application of two different means of financial encumberment for the capacity users, based on identical cost, revenue forecast, and operating overhead factors.

The analysis shows that long term user financing of the international submarine cable around Africa has a revenue disadvantage in excess of $330 million over the life of the cable, when compared to the financing and construction of the international facility by FLAG. It should also be noted here that the African telephone companies' forecast projection represents circuits engaged which account for only 27.90% of the cable around Africa's design capacity.

This analysis points out the advantage of FLAG's financing and constructing the cable over the traditional method of vendor subsidies coupled with long-term financing. When the users finance the international facility, they must bear the cost of inventorying circuits [in this case, 72.10% of the design capacity] which they don't have sufficient traffic to justify engaging. With FLAG funding and building the international facility, financial obligations extend only to those circuits actually required by the African telecom traffic. FLAG assumes the risk of selling the additional circuits above the traffic forecasted in return for a profit. Thus, FLAG's profit motive clearly provides the lowest cost solution [resulting in more than $330 million in additional revenue] for the telephone companies utilizing the circuits of the pan-African cable.

From the donor agencies' point of view, an international facility costing $800 million in new long-term obligations, regardless of the ancillary collateral involved from vendors, ECAs, etc., can only be justified by substantially more traffic than a 27.90% utilization level — or 100%-plus collateral from creditworthy sources. With FLAG funding and building the international facility, the required financial obligations are tied most directly to the actual capacity required by traffic forecast. Furthermore, through the non-recourse collateralization of these circuits, their connectivity requirements, and the shared telecom facilities which drive their requirement — the FLAG alternative offers new commercial leverage to the donor agencies' infrastructure investment in these African telecommunications networks.

QUESTIONS ABOUT VENDOR FINANCING

These are sales incentives for an equipment manufacturer's marketing programme — the home government aligns itself with the manufacturer to protect jobs and exports on the home front. If anything they can aid in smoothing the process of project financing for the international facility. It should be noted that they cannot take the place of the ultimate financial obligation of funding the manufacture and implementation of the international facility itself. These alternative vendor incentive packages actually cancel themselves out in a global competitive procurement process. However, as many in Africa have learned — when applied in individual country marketing efforts where they have little or no competition, these incentive packages have been used as the 'honey' which lures the vendor equipment migration path and ongoing support requirements captured and retained by compatibility issues.

WHAT ABOUT THE MULTINATIONAL CORPORATIONS PROVIDING PRE-CONSTRUCTION CAPACITY COMMITMENTS THAT WILL AID IN THE FINANCING?

AT&T has proposed "a concept tailored to the realities of the global marketplace ... [whereby] ... multinational corporations support the construction of a regional network, in partnership with the telecommunications authorities of the region. They may do this through contractual commitment to use, through subsidies, or through financing arrangements."

The business user represents the most valuable segment of the telecommunications marketplace, even in Africa. The administrative costs and effort required to service the individual residence is a larger percentage of the overall revenue potential of that line than a business' private exchange (PBX) which generates revenues for the firm as it is utilized. The business user has a better ability to pay and requires a larger range of services than the residential component of the telecommunications market universe. As the Egyptian national operator ARENTO has taught us, the cross-subsidization of the overall network by prioritizing business users needs for a premium fee which funds the residential backlog queue is only possible if the business user is serviced within the network and at a fair price.

Let us examine the $800 million international facility circumnavigating the continent of Africa; How can we understand how to bridge the gap between the 27.90% capacity requirement forecast and the 72.10% design capacity inventory? The only possible source of such a prodigious growth requirement is business communications.

When we view the systemic under-employment in Europe, the Pacific Rim and in the Americas, it becomes obvious that the great dislocations of population in search of employment opportunities are now only in the past. Today, it is far more cost-effective and the time to market so much shorter, when a business projects its experience, market information, and close management support into a market target of opportunity through modern wideband telecommunications.

WIDEBAND BUSINESS COMMUNICATIONS WILL BE THE PRINCIPAL ENGINE OF GROWTH, AND THE GREATEST SINGLE SOURCE OF PROFITABILITY FOR TELECOMMUNICATIONS COMPANIES OVER THE NEXT 30 YEARS.

If a major new international facility around Africa is financed through pre-construction capacity commitments from multinational corporations, the overall financial viability of the facility will be jeopardized, and the ultimate profitability of long-distance connectivity for the African telephone company will be significantly reduced.

Of course, if such multinational business users' pre-construction capacity commitments represented the only recourse available, then such a radical diminution of its international telecom franchise might be justifiably contemplated by an African economy bent on obtaining access to the global economy.

However, that is not the case: FLAG proposes to link African telephone companies to the global economy as the lowest cost solution possible — first through connectivity off satellites and onto the FLAG primary configuration, and secondly through a new cable around Africa. A two stage approach offers a more graduated opportunity for African telephone companies to grow new correspondent relationships, targeting business communications from small to medium-sized joint ventures between foreign and local firms — closely supported by a shared telecom facility.

COMPETITION AMONG INTERNATIONAL CARRIERS

FLAG has fashioned an offering of technology and connectivity to enable the African telephone company to compete in its chosen marketplace with anyone, regardless of their size or resources. FLAG's approach encourages the attraction of multiple carriers within a particular country to connect with the African telephone company, thus creating a healthy competition for enhanced access to the development opportunity represented by the African country. Furthermore, by encouraging multiple carriers from multiple countries, FLAG's approach affords more pricing discretion and competitive choice to the African telephone company.

If we examine the current pricing environment and aggressive pricing strategy of an archetypical multinational corporation, this behavior will be much more readily understandable.

DATACOMMUNICATIONS, SEPT., 1992 - Global networking, Eastern style - "... Hongkongbank leases lines from some 38 PTTs for its global network. Its worldwide average monthly price per kbit/s of leased digital lines is down sharply from the $800 per kbit/s per month it used to pay for analog lines."

Exhibit 1

Price per kbit/s of leased lines

UK to Continental Europe: $70-90 / kbit/s / month
Annual Leasing Cost for a 64 Kb/s circuit: $53,760 - $69,120

Transatlantic routes: $100 / kbit/s / month
Annual Leasing Cost for a 64 Kb/s circuit: $76,800

Asia/Pacific regional routes: $150-$160 / kbit/s / month
Annual Leasing Cost for a 64 Kb/s circuit: $115,200 - $122,880

According to large business users, even though leased lines have fallen tremendously in the past five years, there is more room for prices to fall. The cost to carriers of providing these circuits has dropped even faster as a result of spectacular advances in fiberoptic technology. PTTs are realizing substantial profits on international leased lines.

The price score in Exhibit 2 below reflects the average monthly cost to the Hong Kong based bank of transmitting 1 kbit/s on circuits of 64 kbit/s or higher over 200 kilometers. The list is merely exemplary and not a world average.

Contrast the revenue potential of Hongkongbank with a one time payment for 25 years of less than $13,000 per 64 Kb/s, which is the FLAG price per circuit. The revenue loss to the telephone companies on a per circuit basis would look something like the final column at the right of Exhibit 2.

Exhibit 2: Revenue losses telephone companies when FLAG is used

Carrier	Cost	64 Kb/s 1/Mo	200 KM/Yr	10,000 KM/Yr	Revenue loss
Hong Kong Intl	$1.30	$85	$1,010	$50,650	$1,254,000
IDB Worldcom, US	$0.85	$56	$670	$33,400	$822,000
Mercury, UK	$1.90	$122	$1,460	$73,000	$1,800,000
KDD, Japan	$3.90	$250	$3,000	$150,000	$3,700,000
AOTC, Australia	$3.50	$225	$2,700	$135,200	$3,370,000
Teleglobe Canada	$1.20	$77	$930	$46,000	$1,150,000
MCI, US	$0.90	$57	$675	$34,000	$832,000
Singapore Telecom	$6.50	$415	$5,000	$250,000	$6,200,000
BTI, UK	$15.60	$1,000	$12,000	$600,000	$15,000,000
France Telecom	$22.50	$1,445	$17,300	$870,000	$21,700,000
Swiss PTT	$10.30	$660	$7,900	$396,000	$9,900,000

It is readily understandable why multinational corporations would jump at the chance of helping to finance a cable around Africa in light of these per circuit savings.

On the other hand, rather than offering vendor subsidies and advantages such as ECA loans, AT&T could become most valuable as a vendor to the project with a simple commitment to negotiate XX% of the circuits to each country once the cable was completed, on behalf of its WORLDREACH clients. A reasonable discount could easily be derived without forcing the African telecoms to abdicate prospects for profitability through a 'take-it-or-leave-it' offering they are used to seeing from foreign firms bearing gifts.

HOW DOES THE SHARED TELECOM FACILITY APPROACH PROVIDE A VIABLE OPTION?

FLAG's graduated approach begins with bearer circuit connections on its initial fiberoptic facility. To appreciate the values, examine the following Telecom Connectivity Plan derived by ONATEL, Burundi at a recent FLAG data gathering meeting in Bangkok (Thailand).

A.   ONATEL would downlink one E-1 [2.048 Mb/s] via INTELSAT on an IDR carrier to Italcable, Italy.

B.   This 2.048 Mb/s channel would be compressed using digital circuit multiplication equipment [DCME] to yield 120 circuits.

C.   These 120 circuits would then be routed via bearer circuits acquired through Italcable as follows:

   1.   30 bearer circuits to enter the FLAG Cable System at Palermo, Italy and come off the FLAG Cable System at Penzeance, UK, transiting to PTAT Cable System and then on to terminate finally in the USA for a multiple of carriers [MCI, Sprint, AT&T, IDB Worldcom, etc.]

   2.   30 bearer circuits to enter the FLAG Cable System at Palermo, Italy and come off the FLAG Cable System at Fujeirah, UAE. The UAE would be utilized as a terminal party by ONATEL, routing sub-sets of these 30 bearer circuits to Oman, Qatar, and Saudi Arabia, along with circuits to terminate in the UAE as well.

   3.   30 bearer circuits to enter the FLAG Cable System at Palermo, Italy and come oft the FLAG Cable System at Miura, Japan. IDC, Japan would be utilized as a terminal party by ONATEL, routing sub-sets of these 30 bearer circuits to Thailand, Hong Kong, Australia, etc.

   4.   The final 30 bearer circuits would be terminated in Italy for connecting to Italy as well as other European countries, e.g., France, U.K., Sweden, Norway, Denmark, etc.

D.   ONATEL is 'socializing' this plan with four countries of the subregion, Uganda, Kenya, Rwanda, and Tanzania. The objective would be to provide multiples of the above scenario to from the East African countries to common terminal parties to achieve the lowest aggregation prices possible.

E.   ONATEL's strategic objective is to achieve 16 to 20 new permanent facilities with new correspondents with a single E-1 downlink to the FLAG cable system.

F.   With a $1,134 per year charge per bearer circuit for transiting to Italy, the savings per bearer circuit of a satellite / FLAG facility over an all satellite facility to each of the proposed correspondents would amount to more than $90,000 per bearer circuit over the life of the FLAG cable system.

Far more than a transition, the first phase of the FLAG strategy would provide highly cost effective facilities which will provide the most reliable connectivity possible. These new facilities will form the basic relationship platforms which FLAG and the African telephone company will use to market greater business communications connectivity for a 'shared telecom facility'.

Financial analysis of the shared telecom facility demonstrates the ability to service non-recourse debt, while at the same time a substantive revenue stream for the African telephone company is generated.

Using computer and video based applications, the telephone company will be able to concentrate international connectivity services in a single facility which is directly connected to a new fiberoptic international gateway. The call completion rate will be above 90% on this turnkey system. This facility will act as a giant magnet for international calls as foreign firms project their experience, market advantage, new skill-sets and close management support to the joint venture with a local African company which is made possible by the existence and availability of this shared telecom facility.

In each of the 20 countries visited by myself, there are key business leaders who have expressed a commitment to utilizing the shared telecom facility, and in many instances, have offered to provide equity investment in these facilities as well.

As can be seen in this analysis, the high efficiency of this shared telecom facility design provide ample opportunities to vary the offering to ensure global accessibility, cost-effectiveness, and high levels of utilization. However, even at the 5% utilization level, [that equates to 512 minutes per facility bearer circuit per month], a positive revenue stream [$12 million-plus] is generated while the full non-recourse loan is serviced at 8% interest well within 9-1/2 years.

CONCLUSION

The manufacturing and engineering capabilities of Alcatel and AT&T can be marshaled by a FLAG private investment programme. The final design specification can be jointly derived between the FLAG engineering staff and those of the vendors. AT&T's WORLDREACH consortium can collectively negotiate with each African country's telephone company for international access to its marketplace. Alcatel may use its influence [and that of its government] with France Telecom and FT's correspondents, to negotiate for capacity on the African fiberoptic facility in bi-lateral agreements with the African telephone companies it makes accessible to the global telecommunications network.

The shared telecom facility in each African country can derive its telephone company's competitive autonomy through its efficient non-recourse debt service capabilities. The shared telecom facility ability to attract a diversified clientele of small to medium-sized foreign firms seeking to globalize their growth opportunities within Africa will allow African economies to grow jobs and prosperity on African terms.

Regional credit facilities covering the international half-circuits, the landing stations, connectivity thereto, and the shared telecom facilities of 6 to 10 landing points will be syndicated in the commercial lending marketplace. One of FLAG's investors has indicated a willingness to participate in such facilities, providing they are adequately collateralized with a participation in the accounting rate settlements of the circuits financed. Construction of the cable will be done through two simultaneous tranche / construction efforts — one down the east coast and the other down the west coast — continuing until they circumnavigate the entire continent.

FLAG works in perfect harmony with the panafrican satellite system RASCOM. Connectivity to inland countries will be initially accomplished through RASCOM initially, with such transiting circuits financed through the non-recourse structure of the regional credit facilities described previously. Ultimately, however, as broadband requirements expand, and new circuits are required, RASCOM's space segment capacity must be augmented by fiberoptic backbone networks, also financed through the non-recourse structure of secondary non-recourse offerings backed up by the new circuits being acquired.

There can be no greater leverage available from the private sector for the donor agencies' long term financing objectives.

OoOoOoO

APPENDIX 1: TYPICAL ROLLOUT OF THE SHARED FACILITY

PURPOSE AND ASSUMPTIONS

This financial analysis is not intended as a prediction of profitability for the shared telecom facility. Rather, it is designed to reveal the minimum levels of efficiency required for a reasonable amortization period for the resources required to fund and implement the facility. Furthermore, the analysis is structured to provide a working tool with which to evaluate different pricing and operating strategies through a clear delineation of the financial interdependencies of the variables involved in the operation of the shared telecom facility.

1   The configured shared telecom facility in this financial analysis will have 200 high speed workstations assigned cost of $3,000 each.

2.   Acquisition cost of an 100 Mb/s LAN is assigned a cost of $20,000.

3.   20 Videoconference Workstations have assigned cost of $10,000 each.

4.   30 Distance Learning Stations have an assigned cost of $3,000 each.

5.   Two optical-rewritable jukebox equipped servers have an assigned cost of $10,000 each.

6.   5 high resolution color scanners have an assigned cost of $1,200 ea.

7.   5 fax machines have an assigned cost of $600 each.

8.   50 high-speed modems (56 Kb/s) have an assigned cost of $1,000 ea.

9.   An uninterruptible power source for the entire facility is assigned a cost of $15,000.

10.   A facility generator is assigned a cost of $7,500.

11.   The building and wiring is assigned a cost of $750,000.

12.   Fiberoptic connection to the Terminal Station is estimated at 100 km at $15,000 each.

13.   An STM-1 capacity terminal station is assigned a cost of $1,000,000.

14.   5 E-1 [2.048 Mb/s] international half-circuits are assigned a cost of $1,500,000 each.

15.   Annual Operations and Maintenance Charges for these E-1 circuits are assumed at 4.5% of their capital cost, e.g., $62,500 each or a total of $337,500 annually.

16.   One hundred Bearer Circuits [64 kb/s] will be used primarily for data, voice, and fax transmission. It is assumed therefore, that digital circuit multiplication will be employed at a ratio of 3 to 1 to yield a total of 300 effective connection paths for these 100 64 kb/s circuits.

17.   The remaining 50 bearer circuits of the five E-1s are assumed to be dedicated primarily to video transmission, and as such, do not lend themselves to digital circuit multiplication. For purposes of this analysis, therefore, the total circuit yield of the shared telecom facility will be 350 connection paths.

18.   The international tariff assumption is attenuated to the current average cost of calling any country in Africa from Europe - $3.60. Therefore, African

telephone company's share of each tariff per minute which will be utilized to generate a tariff revenue assumption is $1.80.

19. We will assume that each connection path will have a 40 hour per week window, in keeping with the normal business week applicable in most countries. This assumption yields a 100% utilization level of 10,416 minutes per month for each connection path.

20. Usage fees, the following schedule has been assigned to the various technical facilities available for rental at the facility. These usage fees are separate from and in addition to the tariffs charged for an international connection which may or may not be a function of a particular user session:

- Videoconference Workstation - $20 per hour.
- General Purpose Workstation - $15 per hour.
- Distance Learning Workstations - $25 per hour [These entail the use of the Optical Jukebox servers as well as the workstations in order to function]
- LAN Services, billed on the basis of a user enterprise's assigned LAN sector at the rate of $500 per week, cover costs of resource allocation, profile of station privileges, backup, etc.

21. We have broken Facilities Hardware Utilization and Connection Path Utilization into a graduated series of levels. The number of these levels is to create a comprehensive toolset in this worksheet to allow for the analysis of different levels as well as different combinations of levels of utilization between the Facility's hardware utilization rate and the telephone tariff utilization rates. Furthermore, it will be assumed in this analysis that the ratio of utilization rates between tariffs and facility hardware will be four to one in favor of hardware utilization. This is true because off-line use of equipment based on periodic high-speed downloads and uploads over the international circuits will become the most cost-effective user model. Additionally, this user model can more readily accommodate more enterprise users simply by increasing the number of equipment units available via the local area network.

22. The non-recourse loan is structured as an 8% rate of interest loan whose debt service is tied to a 50% participation in the accounting rate settlements share of the borrower telephone company. In this design, each $1.80 per minute the African telephone company is entitled is divided equally between the non-recourse debt obligation and the African telephone company. Furthermore, the loan is structured to be serviced by 100% of this cash-flow participation during the period of settlement, so that once the outstanding interest accrued to that point has been serviced, any remaining residual revenues resulting from the accounting rate participation will go toward retirement of principal.

23.   It is anticipated that private investors and technology proficient operators may be strategically aligned with the telephone company in the capitalization [both financial and 'in kind'] of the shared telecom facility on some basis that would serve to strengthen the success potential of the shared facility and lower the borrowing requirements of the shared telecom facility venture significantly.

Telecommunications and Development in Africa
B.A. Kiplagat and M.C.M. Werner, Eds.
IOS Press

CHAPTER XIV

ADDRESSING THE TECHNICAL REQUIREMENTS OF AFRICA - AN AFRICAN TESTING ORGANISATION AND ITS IMPACT ON NETWORK QUALITY

Reward Kangai
Director Manufacturing & Projects
PTC Zimbabwe
Kari Vaihia
Consultant, Telecom Ltd
Finland

ABSTRACT

The establishment of an Environmental Tropicalisation and Testing Centre (ETTC) Harare, Zimbabwe, will promote telecommunications equipment manufacturing in the Southern African Development Community (SADC) and result in enhanced network reliability throughout the region.

The demand for the establishment of the center was expressed in several conferences. Based on that demand, UNIDO made a project plan for a regional environmental tropicalisation and testing centre. The government of Finland donated US$ 1.05 million for the project. Posts and Telecommunications Corporation (PTC) of Zimbabwe functioned as a local counterpart and the implementation was done by the telecommunication consulting company Telecon Ltd., subsidiary of Telecom Finland. The first mission took place in September 1991 and the first phase was completed in 1993. The implementation of follow-up phases will depend on future funding for the project.

The results of this initial phase can be utilised in many ways by all interested parties in the SADC countries. The calibration laboratory operates in conformance with the standards set internationally by the ISO guide 25 and calibration is being carried out in premises of controlled environment as stated in the WECC recommendations. All the working standards used are traceable to the primary standards. The following types of calibrations can be carried out in the ETTC calibration laboratory:

1. Low Frequency AC/DC Voltage and Current Measurements
2. RF Power Measurements
3. Oscilloscope Calibrations
4. Frequency Calibrations
5. Low Frequency Impedance Calibrations

The trained environmental group can give advice in many essential environmental issues (protection, material selection, maintenance) encountered in the African environment. They can also carry out some basic evaluations concerning the quality of imported equipment. The environmental tests will be carried out when the next phase of the project is implemented.

The testing center itself will be equipped with all modern facilities (PC LAN, office equipment) to give customers good services and reports of high quality.

The first results show huge demand for calibration of electrical instruments: more or less all instruments calibrated so far were out of specification!

INTRODUCTION

Following the report of the Independent Commission for World Wide Telecommunications Development called "The Missing Link", the following conferences aimed at finding ways of promoting the manufacture of telecommunications equipment in Africa were held:

a   UNIDO/PATU Seminar on prospects for the manufacture of telecommunications equipment in Africa, Harare, Zimbabwe January 6-11, 1986.

b   Third Ordinary Session of the PATU Conference of Plenipotentiaries, Arusha, Tanzania March 1-7, 1986.

c   PANAFTEL seminar/workshop on maintenance between African
        Administrations and Equipment Suppliers, Abidjan, Ivory Coast, February 28,
        1986.

d   ITU/PANAFTEL African Telecommunications Conference, Tunis, Tunisia,
        January 12-16, 1987.

As "the Missing Link" identified the lack of telecommunications manufacturing
facilities in developing countries as one of the key factors for the inadequacy of
telecommunications services in the developing countries, the above conferences were
held with the objective of seeking ways of promoting telecommunications equipment
manufacturing in Africa. The idea of setting up an Environmental Testing and
Tropicalisation Centre for the SADC region came up from these conferences.
Accordingly, the United Nations Industrial Development Organisation (UNIDO) was
tasked to make a project proposal which was produced and approved by UNIDO in
January 1988. Harare, Zimbabwe, was selected as the site for a centre.

THE OBJECTIVES OF THIS PROJECT WERE:

1   To enhance the national and regional awareness of the importance of the
        environmental testing and tropicalisation of telecommunications equipment
        through training programmes and advisory services.

2   To establish a SADC Regional Environmental Testing and Tropicalisation
        Centre (ETTC) for telecommunications equipment, with premises, technical
        facilities, laborato   ries, workshops, equipment, technical library and
        database, trained personnel and management system, to be capable of:

        ❖   carrying out environmental tests to verify the tropic-proofness of
                telecommunications equipment.
        ❖   developing and applying tropicalisation techniques and material to the
                already installed and/or locally designed and/or manufactured
                equipment
        ❖   providing advisory service on the subjects within its field of
                competence.

SADC MANUFACTURING STRATEGY AND ETTC

The Transport and Communications Commission of SADC (Southern African
Transport and Communications Commission, SATCC) commisioned a study aimed at
increasing the manufacture of telecommunications equipment in the region. This study
was executed by HN Engineering Inc. (HNE) of Vancouver, Canada. The study
recommends the following goals which were approved by SADC:

◻   to achieve SADC regional manufacture of 80% of the telephone cables and electrical and mechanical products used by the SADC telecommunications administrations by 1995;

◻   to achieve SADC regional manufacture of 30% of electronic products used by the SADC telecommunications administrations   by the year 2010.

Whilst a number of measures are already in place to ensure the implementation of the above objectives, the realisation of these goals will largely depend on the confidence of the SADC telecommunications administrations on quality aspects of locally manufactured telecommunications products as they (SADC telecomms administration) are the end-users. It is in this regard that the Environmental Testing and Tropicalisation Centre will play a leading role in this confidence building in locally manufactured products, as ETTC will provide the following services:

◻   Selection of imported equipment as well as tropicalisation of locally produced equipment;

◻   Carrying out environmental tests to verify the tropic-proofness of telecommunications equipment;

◻   Carrying out electromagnetic compatibility and electromagnetic interference tests for telecommunications equipment;

◻   Telecommunications test equipment calibration;

◻   Providing advisory service on environmental aspects related to telecommunications equipment e.g. lightning, temperature,   humidity aspects, etc.

The services to be offered by the ETTC will enable quantitative assessment of the quality of telecommunications equipment both locally manufactured and imported and thus enable the equipment users to verify for themselves that the equipment being purchased, will be able to withstand the environmental conditions prevailing in their region.

This will be possible through subjecting prototype equipment to climatic, mechanical, electromagnetic interference testing and checking the performance of the equipment against specifications laid down by SADC telecommunications administrations. In this manner, telecommunications manufacturers will be able to see areas requiring improvements on their equipment, whilst on the other hand, telecommunications administrations will have the confidence that locally manufactured equipment is of high standards.

The local telecommunications and electronics industry will aim at meeting these high standards for their equipment to be acceptable by SADC administrations. Others factors (price, deliveries etc.) being equal, locally manufactured telecommunications equipment meeting SADC specifications will therefore be put on a competitive footing against foreign equipment.

In order to specify more precisely the ETTC functions, information was gathered about the telecommunications manufacturing and manufacturers in SADC countries. The most potential area seems to be the products for the external plants of telecommunications operating companies, like cables, pools, cabins, trays, junctions etc., or in the other words the so called mechanical products.

Electromechanical products are being manufactured or assembled too, for instance in Zimbabwe. However, operating companies are now in a process of digitalising their networks, leaving no place for electromechanical products in future. On the other hand there may be possibilities to produce in some limited scales certain types of PCB's (printed circuit boards).

The main emphasis could be to produce quality inspectors for incoming products manufactured abroad and to help local manufacturers as much as possible in those areas where they already have some capabilities and where they can expand and export.

One problem has been lack of confidence of operating companies in suppliers from other SADC countries. This confidence could be improved by supplying specified and tested products. When testing facilities are lacking, the customers have to rely on the supplier only. Some specifications could be refined so that they would give more information about quality. For instance, finishing is often mentioned without detailing thickness.

OUTPUT OF THE CENTER

The objectives of the ETTC project remained basically unchanged since the beginning, but the planned outputs of the center had to be modified slightly in order to match real world demands with the resources available for the first phase. The rest of the outputs will be taken care of during the following phases. The original outputs were:

- ❖ environmental testing laboratory
- ❖ EMC laboratory (electromagnetic interference compatibility)
- ❖ materials and processes laboratory
- ❖ advisory services
- ❖ administration
- ❖ co-operative programme ETTC-University-Industry

The project was started by establishing an advisory service for environmental issues and calibration laboratory. Even if there would be a need for an EMC laboratory, that output could not be started because the costs to establish this type of center were prohibitive. The establishment of a materials and process laboratory was cancelled at this phase, mainly because those services were available in Zimbabwe.

The core staff of 10 went through a two month training course in Finland. The staff comprised engineers and technicians with different background in telecommunications. The emphasis in the courses in Finland was on hands-on courses and comprised several visits to laboratories and companies. The quality management was a common factor for all courses.

CALIBRATION LABORATORY

The ETTC calibration laboratory has been operational since January 1993. The calibration staff comprising of two engineers and two technicians underwent an intensive 8 week training course in calibration in Finland. At the moment, staff is gaining experience while calibrating test and measuring equipment from PTC Zimbabwe. By the end of 1993 they will be able to serve clients from companies throughout the SADC region.

The laboratory will maintain the standards used in calibrating by having them recalibrated at regular intervals at an approved National Standards Laboratory (e.g. National Calibration Service, South-Africa), thereby maintaining traceability of the measurements.

The laboratory operates in conformance with the standards set internationally by ISO. Quality management is being implemented in the laboratory and calibration is being carried out in controlled environment premises conformant to WECC recommendations.

The following types of calibrations can be carried out in the ETTC calibration laboratory:

1.   Low Frequency and DC Measurements, Current and Voltage Measurements
2.   Power Measurements
3.   Oscilloscope Calibrations
4.   Frequency Calibrations
5.   Low Frequency Impedance Calibrations

The emphasis during the first phase is to gain experience on the low frequency side and then to proceed to higher frequencies. The higher the frequencies, the more expensive the equipment will get. The higher frequencies are also susceptible to many external disturbances and thus need a special disturbance free room.

The calibration group will go in more details in areas such as:

❖   ISO 9000
❖   Electrical standards
❖   Disturbances
❖   Development of calibrating of telecommunications measuring equipment

The first results have shown that the demand for the calibration of electrical instruments is huge: more or less all instruments calibrated so far were out of specification! The instruments seem to function normally but with a wide error margin. When these uncalibrated instruments are used for different measurements and ajustements of networks they can easily generate faulty situations which are very difficult to localise.

ADVISORY SERVICES ENVIRONMENTAL ISSUES

Annex A presents a list of typical problems that operators are likely to meet in Africa. The main problems are lightning rain, dust and intensive sunlight. From the maintenance point of view the wide variety of equipment and materials is a problem. Besides the natural environmental factors, thefts of telecommunications items complicate things further.

From the list it is clear that nature is not homogeneous. The conditions can vary a lot and the equipment can be facing conditions they are not meant for. It is common that equipment is designed and manufactured to work only in a certain "average" condition and when that average condition is exceeded, the failure rate can increase hugely. That is why IEC (International Electrotechnical Commission) has classified the conditions in their publication IEC 721. This document is composed of several sections which define accurately all essential parameters and the severities equipment can meet.

721-1   Classification of environmental parameters and their severities

721-2   Environmental conditions appearing in nature
      721-2-1   Temperature
      721-2-2   Precipitation and wind
      721-2-3   Air pressure
      721-2-4   Solar radiation and temperature
      721-2-5   Dust, sand, salt mist/wind
      721-2-6   Earthquake vibration and shock
      721-2-7   Fauna and flora

721-3   Classification of groups of environmental parameters and their severities
      721-3-0   Introduction
      721-3-1   Storage
      721-3-2   Transportation
      721-3-3   Stationary use, weather protected
      721-3-4   Stationary installation, non-weather protected
      721-3-5   Ground vehicle installations
      721-3-6   Ship environment
      721-3-7   Portable and non-stationary

IEC 721-1 contains a list of those parameters that can have an essential effect on material or performance of the product during its lifetime.

IEC 721-2 contains information about the environmental parameters in nature, their physical background and statistical appearance. Because some parameters have synergistic effects or they belong to the same group they are in the same document.

IEC has not made efforts to create a system for transformation of classified environmental stresses to specified test methods and severities. This task is left to the user of the documents. However, the environmental experts in the Nordic Tele-communications Administration (NT/ENV) have made systematic transformation which is utilised by the European Telecommunications Standards Institute (ETSI).

It should be tested somehow whether a product fulfills the specifications (performance, physical dimensions, functionality) in the environmental classes defined by IEC. For that purpose, about 60 test methods are available in IEC 68. In order to understand the variety of the tests we can mention some typical ones:

68-2-1	Cold
68-2-2	Dry heat
68-2-3	Damp heat, steady state
68-2-6	Vibration
68-2-9	Guidance for solar radition testing
68-2-14	Change of temperature
68-2-17	Sealing
68-2-27	Shock
68-2-29	Bump
68-2-42	Sulphur dioxide test for contacts and connections
68-2-43	Hydrogen sulphide test for contacts and connections
68-2-44	Soldering
68-2-50	Combined cold/vibration (sinusoidal) test for both heat dissipating and non-heat-dissipating specimens
68-2-52	Salt mist, cyclic

Some of the tests are very simple to do but some of them (especially vibration tests) require very expensive testing equipment. Besides that, the room where the test will be accomplished may require environment conditions (temperature, humidity, cleanliness, noise). That is one obvious reason why it is sensible to concentrate some of the testing facilities. Concentration is further suggested because of the special competence on the use of the environmental classes, selection of the test and interpretation of the test results required.

The classification system is really a good thing which can be utilised by manufacturers. The best results however will be gained when the classes are chosen together with the potential customers. In that situation one can decide whether the class must be changed or if equipment needs some extra protection. Also the tests to verify equipment performance can be chosen in close cooperation among both sides.

An advisory group is dealing with issues like

- ❖ Environmental classification
- ❖ Design & Planning requirements of premises
- ❖ Corrosion & surface finishing
- ❖ Quality inspection of telecommunications products

The environmental group has found that where quality is concerned the problem is very often the lack of proper measuring tools. Sometimes even very simple tools could be enough to screen out most of the substandard products. The group has got training to accomplish visual checks on PCB's, which can be verified with some helping tools like a magnifier and microscope. They can also use several reference books to compare the manufactured products with products with approved quality.

CONCLUSION

ETTC is specialised for calibration of telecommunications measuring instruments and environmental testing of telecommunications equipment. The center has been designed with African requirements in mind.

Once completed, the center will be unique in the SADC region. At the moment the center can calibrate the electrical instruments against reference standards according to the international specifications. Besides this, the center can help in environmental issues, which telecommunications equipment is likely to meet in the field.

Once the next phases are completed, the center can offer very versatile calibration services and accomplish environmental tests according to the IEC 68 specifications.

One of the goals for the center is to be an accredited calibration laboratory for certain electrical measurements by 1998.

ANNEX A : POSTS & TELECOMMUNICATIONS CORP. (ZIMBABWE) REPORTS FEB-APRIL 1993

1. Vandalism (joint box); e.g. kids cutting wire for toy making
2. serious shortage of MAPL and DEL phones for maintenance
3. cable thefts in many areas
   - ❖ 16 pr. open wire at Mandley (Mutare rd) stolen 4 times in a week
   - ❖ cable thefts in manholes
4. only some of PCO's are working
5. solar panels replaced and farmers providing guards for protection
   - ❖ stolen solar panels
6. bushcut

7.   transportation is a problem
     ❖   vehicle on the road 0 - 100%
8.   power problems
     ❖   new charge controllers needed
9.   regenerator housing damaged due to corrosion
10. cable feeding damaged by contractor
11. materials in short supply
     ❖   soldering flux, telephones, insulation tape, staples, spindles, cells,
         salamoniac blocks
     ❖   APL telephones, rare gas tubes, indoor wire, protector pole mounting
12. heavy rains increase the number of faults
13. shortage of spare phones and problems compounded by long turnaround time
     for faulty phones sent for repairs
14. centre core of coaxial cable dry jointed
15. faulty batteries
16. rectifier problem (packed up thyristors etc.)
17. reliability records (Mazoe earth station)
     ❖   reliability = total trouble time/time
     ❖   trouble time = number of affected channels/total number of channels
18. commercial power failures
     ❖   abnormal voltage, open phase, power surge
19. cable faults; cleaning takes time due to lack of specialised equipment
20. some posts are hard to fill because of housing problems
21. battery charging problem due to a faulty solar power regulator
22. lightning damages
23. faulty PCO's
24. stolen batteries
25. faults cleared on average (%) same day 72, next day 85,7 days 95, 14 days 96,
     30 days 97
26. corrosion of cables
27. holes in cables caused by ? (corrosion, insects)
28. damages of materials by improper storage conditions
29. cooling capacity of a exchange room insufficient; malfunctions

CHAPTER XV

TELECOMMUNICATION EQUIPMENT PROCUREMENT STRATEGIES FOR AFRICAN NETWORK OPERATORS

Geoff Hainebach
Joint Managing Director
Siemens Limited
South Africa

There is little or no controversy about the importance of telecommunications as one of the essential pre-requirements for economic development. The others are: Water Supply; Effluent Control; Energy Supply; Transport. Health and Education are often quoted in the same context but are dependent on the previous pre-requirements. Small wonder then that most African countries have not developed as hoped and required.

Statistics show that Africa is by far the worst continent in terms of telecommunications development.

Virtually all telecommunications resources in African countries are owned and operated by a government agency, often reporting directly to a government department. Their budgets still are treated as a typical state budget. Surpluses are returned to the treasury and deficits funded from the exchequer.

CONTRIBUTION TO THE PROBLEM

Most African Telecommunications Network Operators feel obliged to fund new projects with soft loans or aid finance. In contrast to the views of the World Bank which believes that funds for telecommunications development can to a large extent be generated internally or secured from internal cash flows, African Telcos either have no access to such funds because they are obliged to return surpluses to the exchequer, or because often the projects are large in comparison to the available cash flows.

Soft loans or grant aid finance usually comes in two ways. If the source is neutral, such as multilateral institutes as the African Development Bank or the World Bank, the project is subject to audit by specialised consultants. The studies conducted by these consultants generally take from 6 to 24 months and are very expensive, as these consultants are vitually always expatriates. Cases of 10 to 25 percent of final project value spent on such consultancy are on record. The reports are often extremely detailed and cover virtually every aspect that can be imagined. In the end, funds are seldom available to cover the consultants' recommendations.

Such detailed studies require considerable expertise to understand and critically review, expertise that is often not available in the country applying for the finance, nor in the financial institution considering granting the finance. The experience and know-how gained in doing the study is gained by the consultants and does not remain with the organisation in the country where it is needed.

It is therefore the consultants who are usually asked to implement their recommendations, but because of limitations in the availability of finance have to do so against reduced scope and specifications.

Tenders are called for and suppliers squeezed to the maximum. The winning tender is therefore stripped of all items not immediately required for startup. Among elements that risk being supressed are sufficient maintenance training, adequate spare parts inventory, sufficient storage for future software upgrades, or rack space for systems expansion. Whatever is supplied, is supplied on a minimum basis, too often meaning that little or no support is included.

After the commissioning tests are completed, the installation crew returns to its country of origin, seldom if ever to return, simply because even if they are required, the funds to get them back may not be available. If local technicians have been trained under the contract, they seldom are given the opportunity to acquire the necessary experience required to deal with serious problems, especially if these relate to the network as a whole, rather than just the specific equipment on which they have been trained.

The other option is to use aid funds. These can be soft loans or grants and are either conditional to supply from a vendor in the donor country, or free from such ties. In the case of unconditional aid, the funds are mostly disposed as described above by neutral financial institutions. If conditional, the aid comes in the form of equipment, not money.

The donor countries naturally try to obtain the best public relations results from their generosity and rarely focus their aid on telecommunications in the same country during successive financial periods.

All the financing methods as mentioned above have but one effect. Every single project in each beneficiary country results in the acquisition and installation of equipment or technology from a different source. The ensuing logistics are horrifying and the main reason why the state of repair and functionality in African telecommunication networks is generally so poor.

Not only is the typical African national network small, comparable with a private corporate network in an industrial country. In many cases, only a small proportion of the lines connected to these public networks are operational. Examples on record include countries with a network with no two switches identical. The situation with transmission equipment is not quite so bad due to the relative ease with which dissimilar transmission systems can be matched in a network, but nevertheless the picture of logistical trouble is not much better either.

PROCUREMENT VERSUS OTHER COSTS

Procurements costs are not the only costs to be considered in telecommunications projects. For equipment in public communication networks, they represent only 20 percent of the total on average. Operation and maintenance, as well as administration costs represent the rest in equal proportions, i.e. 40 percent each.

In the cases described above which apply to the majority of African countries, operation and maintenance costs as well as administration costs would thus be much higher, but cannot be afforded, resulting in severly under-performing networks which yield a fraction of the revenue for which they were designed.

A typical response is to increase tariffs making telecommunications even less affordable to an already financially stressed population. At the same time, decreased operational efficiency is making telecommunications even less accessible.

TOWARDS A MORE RATIONAL APPROACH

At the present stage of development, few African Telcommunications Operators can afford to support more than one supplier's technology, but for political cum business reasons are not at liberty to support less than two. Therefore, supply contracts should not be for equipment alone, but also include sustained availability, whereby the supplier is paid according to the functioning of the equipment he supplies and maintains. Until the operator is fully capable of operating and maintaining his network, the supplier should be contracted to do this and be remunerated by revenues derived from only the part of the network for which he has a responsibility. For example, the supplier would receive x percent of revenues for a period of y years. The values of x and y depend on the traffic for which the network was designed. This may appear to result in a higher cost per line unit. However, it will be more economical than the present approach which is detached from any notion of self-sustainability of installations and therefore result in higher revenues per line unit, driving the penetration upward. Perhaps this approach could even reach the World Bank's ideal of a self-funding network.

In addition, it would make economic sense to combine the resources of as many Telecommunications Operators as possible to form a co-operative planning, engineering and purchasing organisation, in much the same way as Bellcore does for the American Regional Bell Operating Companies (RBOCs), rationalising the expensive functions and increasing buying power. In addition, the experience and know-how gained would remain in the region and contribute to the advancement of its telecommunications sector. The Southern African Transport and Telecommunications Committee (SATTC) would be an ideally placed vehicle for such co-operation, especially if the relatively large national operating company Telkom of South Africa were to be a party.

Achieving this degree of co-operation would require the sacrifice of a certain amount of political control and autonomy. However, the benefits to the different governments would far exceed the sacrifice of control and possible subsidy to the exchequer via the facilitation of economic activity and the taxes gained directly and indirectly therefrom.

PART IV

NEW TECHNOLOGY

Telecommunications and Development in Africa
B.A. Kiplagat and M.C.M. Werner, Eds.
IOS Press

CHAPTER XVI

ADVANCED TECHNOLOGY CONCEPTS AND THEIR SUITABILITY FOR DEVELOPING COUNTRIES

Reginald R. Teesdale
Managing Director Sestel SA
France

Designers of telecommunications networks are offered a rapidly increasing diversity of techniques and technologies and those in developing countries face a major challenge when seeking an optimum strategy for up- grading their networks.

This paper discusses some of the basic issues involved in making major strategic decisions and attempts to show how modern technology can best be deployed in developing countries.

'REGIONAL' AND 'GLOBAL' SOLUTIONS

Although much attention is focussed on technology, we have to bear in mind from the outset that technology is a means to an end, a method of implementing a design and not the design itself. We therefore begin with the factors that have an impact on the design of the network.

Every region has its own characteristics that affect the network design and the choice of appropriate telecommunications strategies and technologies. In particular, the traffic flow, calling pattern, communities of interest and terrain may vary widely from region to region. This would tend to suggest that every telecommunications network is unique, and this claim is in fact frequently made by the designers of private networks.

The 'uniqueness', however, may lie at a tactical level rather than a strategic level and there are good reasons to look for global solutions and to regard any given region as part of an evolving global network and not as an isolated 'island'.

One reason is cost. Modern technology tends to have a price relationship that is relatively insensitive to complexity but sensitive to volume so that complex components cost no more that simple devices, but must be widely used to be affordable. For example, high-technology complex devices are now frequently found in mass-market low-priced products such as video tape recorders and compact disk players. A network that uses the same devices as the global network is likely to enjoy low capital costs and continued manufacturing support.

The second reason is interworking. The new network must interwork with the existing global network and large multinational business users need to integrate their factories, workshops and offices, located in the developing country, into their own in-house network and will therefore need compatible networking and technological solutions.

Interworking is essential at the level of the core network and at the enterprise network level, but it is also highly desirable at the level of the individul residential or business subscriber. In particular, the originating and terminating facilities and special services should operate in the same manner wherever the caller or called party is located. Any telephone call, no matter where it originates, may now find itself routed to an answering machine, voice-mail terminal, transfer number etc., or, if mis-dialled, may terminate on a Fax or modem line. Users increasingly need to be familiar with such eventualities or they will be afraid to use the telephone.

There are therefore good reasons to avoid 'special' region-specific or country-specific solutions but even the developed countries design 'stripped down' versions of their standard equipment for special areas such as sparsely-populated regions and such needs often affect the modularity of the base product. Developing countries should therefore seek out mass-useage products that have an appropriate modularity to enable effective custom-engineering and special country-specific solutions should only be envisaged as a last resort.

In the past, countries and regions often designed 'their own equipment', partly due to the NIH-syndrome (Not Invented Here), partly to develop in-house skills and partly to ensure indegenous manufacture, but the complexity and development cost of modern equipment makes such initiatives increasingly unattractive. NIH is no longer an admissible attitude, system skills can be acquired through custom-engineering and local manufacture should be a contractual obligation.

THE DANGER OF 'SHORT-CUTS'

TECHNOLOGY SHORT-CUTS

As we have seen, there is a need to cater for regional differences within a framework defined largely by the global marketplace. This duality of purposes makes short-cuts and axioms very dangerous, especially when the axioms hide a deeper reality. For example, ATM (Asynchronous Transfer Mode) is frequently linked with broadband and one possible axiom may be "we need broadband so we must have ATM" - or vice versa, but ATM is in fact a bandwidth-independent technique and the raw bandwidth derives from the layers below. ATM, however, enables flexible useage of the raw bandwidth and so enables service-independent networks and bandwidth-on-demand. Such features are often potentially much more valuable than broadband itself.

New developments often have very evident attributes, such as bandwidth, and become associated uniquely with them, but their real value to a developing country may lie elsewhere and relate to future-proofing, procurement issues, spares provisioning, the skill-mix needed to operate the system and so on. Fibre-optic cables offer a wide raw bandwidth, for example, but they also offer immunity against induced noise and it is this second characteristic that enables the whole error-control structure to be re-thought when fibre is employed.

SHORT-CUTS RELATING TO THE LOCAL ENVIRONMENT

We also need to beware of axioms relating to the local environment. The availability of a large moderately-skilled workforce, for example, can be used as an argument to install a 'simple' manpower-intensive system that will provide employment. But older systems tend to be very unreliable compared with modern 'maintenance-free' systems and a large but bored workforce tends to interfere with the equipment when it is working fine and cause further outages.

Modern technology is based on the idea of unit replacement and off-line repair, rather than component replacement in situ. This means that a developing country can establish repair centres and concentrate skilled manpower in a number of well-equipped workshops that can rapidly become 'centres of excellence', rather than rely on a widely dispersed maintenance workforce with varying levels of training.

SHORT-CUTS RELATING TO 'SIMPLICITY'

In any case, pleas for 'simple systems' tend to duck away from the follow-on question, 'what do you mean by simple?'. In many ways, the older equipment was exremely complex, especially for the maintenance technician, and no-one who has had to interpret fault printouts from electromechanical equipment will willingly forego the computer that explains the problem in plain language. The complexity of modern systems tends to be locked away in the chips and the system software and the man-machine interface is much simpler than it used to be.

A modern system is therefore much simpler to administer and maintain but the critical complexities, especially in the software, have to be very well protected.

SHORT-CUTS RELATING TO SYSTEMS

We are all aware of a number of telecommunications networks that have developed over time, and it is tempting to use short-cut decisions and 'opt for network ... x'. But we should not choose an analogue PSTN (Public Switched Telephone Network), digital PSTN, PSDN (Packet Switched Data Network), NB-ISDN (Narrow Band Integrated Services Digital Network), frame relay or B-ISDN (Broadband Integrated Services Digital Network), or make a quick decision to use radio rather than cabled networks or vice versa, in an attempt to simplify the problem, because this apparently simplifying decision may in fact cause unsuspected complexities elsewhere. A PSTN may well provide an excellent local telephone service but will never interoperate fully with the global and enterprise integrated services networks. The NB-ISDN cannot provide for high-speed data networking or quality images and the B-ISDN does not exist. Radio works best when it is allied with an extensive cabled network. Most cellular radio calls spend most of their time in the cabled PSTN. Cable often works best in conjunction with radio distribution systems. There is no real alternative to a searching analysis of requirements and options.

SHORT-CUTS RELATING TO STRATEGY AND IMPLEMENTATION

Short-cuts also tend to mix technology and implementation with strategy and networking: 'we will use an ISDN with PDH/ATM'. However, it is essential that these are treated separately. Whatever network/system solution is adopted, technology will continue to evolve and the objective should be to enable the use of evolving technology without major changes to the basic system infrastructure.

The basic enduring decisions relate to the system and network design: all technology decisions are in essence temporary and they all become out-dated sooner or later.

INFRASTRUCTURE DESIGN AND IMPLEMENTATION

Some of the most sweeping changes that have taken place within the last few decades have separated infrastructure design from implementation.

In analogue switching systems, the network design tend to be embodied in the components and the interconnect wiring so that a significant change implied an enormous amount of hard-wired modification to individual exchanges. Modern software-based systems are designed to enable different generations of equipment to interoperate within the same system and most changes involve only software changes (although these lead to very complex problems of software configuration control).

These conceptual changes have so altered our way of thinking that even if we now went back to the analogue PSTN, we would almost certainly design it differently. We can now see that many of the old 'axioms' were in fact assumptions that were inadequately questioned. For the old-time telephone engineer, the Stronger system (or the crossbar or rotary system) was 'the telephone system', not just a specific implementation of it. The advent of common-channel signalling and processor control made it impossible to 'return to the past'. In particular, the advent of processor control required telecommunications engineers for the first time to think about the telecommunications process in an implementation-independent way and the insights that have arisen from this re-think cannot now be forgotten or ignored.

THE EVOLUTION OF PUBLIC SWITCHED NETWORKS

The philosophy adopted in the early days of telephony was to 'bring a telephone into every household'. Telephony was seen as a commodity like clean water that should be made available to all citizens, but its impact on the efficiency of enterprises was largely unremarked.

From early time, however, business users could rent lines and equip their own switching plant and the provision of leased lines in fact marked the first breach in the idea of a universal switched telephone service. This breach was unremarked because the private telephone exchanges were simply small public exchanges with a few added facilities.

Digitalisation of the PSTN improved reliability and quality and considerably reduced both the capital cost and cost-of-ownership of the PSTN, but the general aim was still a pervasive public switched network. Digital leased lines were provided and digital private exchanges began to differ from public switches as technology evolved and business users demanded more sophisticated facilities. The private PABX became, not a cut-down version of a local exchange, but a very sophisticated small switching system that evolved much more quickly than its public counterpart, and many modern small digital public exchanges are in fact descendants of private exchanges. They particularly provided an impetus towards the design of small memory-efficient control processors and cheap memory, since the early machines imposed a major start-up cost burden on the exchange.

At the same time, modern data services over the PSTN became increasingly unacceptable due to reliability, noise and bandwidth problems and both circuit-switched and packet-switched data network were initiated.

The 64 Kbit/sec ISDN came about by adding new features to the digital PSTN to provide for low-speed data services and to integrate circuit-switched data into an integrated system, along with circuit-switched low-speed video and image services. Packet-switched data networks continued to develop separately according to the X.25 protocols. The same digital interconnect backbone was used for leased lines and for all circuit-switched, packet-switched and channel switched services up to 2 Mbit/sec. and these were segregated only near to the terminal. ISPABX's provided ISDN switched services within private networks using digital leased lines.

Gradually, the focus was shifting away from universal customer service towards universal electronic interconnect highways serving terminal nodes that are customised for different types of useage or service (circuit-packet, broadband-narrowband, broadcast-point-to-point etc.) and the 'universal telephone service' was becoming just one of many services carried over the same infrastructure. The explosion of cellular radio suddenly re-focussed attention upon telephony, which had come to be regarded as a rather moribund and uninteresting service, and considerably increased the telephony traffic and signalling traffic over the wired telephone network. But by now data traffic was beginning to become a significant fraction of the total carried traffic and was no longer the 'poor relation' of telephony so that data considerations played an increasingly important role in telecommunications network design.

The advent of processor control and common-channel signalling caused telephony engineers to re-consider their designs and they chose to utilise the emerging ISO 7-layer model and have considerably extended it to cover all types of service. These new insights led to complete separation of the control infrastructure from the transmission/switching network and bandwidth became an attribute of the switching/transmission layer, not of the system.

The B-ISDN was a further development of the ISDN, using the same control infrastructure and the SS No 7, but with a wideband switching/transmission overlay, and was intended to provide ISDN plus universal wideband service at 150/600 Mbit/sec. At the time of its inception, it was generally accepted that the transmission/switching or transport segment would be based as usual on fixed channelling, but asynchronous techniques had meanwhile come to fruition and almost the first major decision taken by the ITU-T (then CCITT) organisation was to base the B-ISDN transport on ATM.

The timescale for the B-ISDN as a fully-switched public service has moved progressively further away and few now believe in the realisation of this global concept in the near future. The real demand for broadband user-visible services is still weak and is eroded by improvements in image and video coding that enable theseservices to exist in the ISDN. The region 2-50 Mbit/sec. is increasingly exploited to handle medium-quality image and video and only a few niche applications require more than 100 Mbit/sec. The cost of the 'last kilometre' of fibre cable is still a major obstacle and broadband radio suffers from lack of brandwidth unless very high carrier frequencies are employed (which create their own problems). The standards process, both within ITU and ISO, is long and tortous and pressure groups such as the ATM Forum now create ad hoc agreements that are widely regarded as de facto standards, so that there is no single authoritative standards body.

Above all, the widespread liberalisation and privatisation of the telecommunications business has destroyed the old comfortable if slow-moving development process and replaced it by a commercially-driven and somewhat anarchic process that leads to piecemeal improvements rather than to global strategies. Telecommunications has become much more 'tactical' and less 'strategic' in the last decade and suffers more than its fair share of commercial hype.

The net result is that ATM as a concept is competing successfully with similar ISDN-based services such as frame relay, commercial services such as SMDS, which closely resembles ATM, are now widely offered and bandwidths are edging upwards towards 34/50 Mbit/sec. Broadband is being introduced piecemeal, in a rather ad hoc unplanned maner, but enthusiastically, in most of the developed countries. Meanwhile, a major shake-out of suppliers is likely within the next five years.

DEVELOPING COUNTRIES - LEAPFROG OR GO-SLOW ?

In most developing countries the target of a 'telephone in every household' is still very much a dream and there is a natural reaction to call for 'a no-frills universal basic telephone service' and to worry about up-dating later on. In other words, for the developing regions to step through the same process as that outlined above.

There are a number of counter-arguments to this. As shown above, times have changed, technology has changed and concepts have changed: we cannot 'turn bach the clock', forget what we know about system design, common channel signalling, computer-assisted network management and processor control, and build another analogue PSTN. In any case, a 'no-frills' telephone service would be unlikely to cost significantly less than a modern telephone service, at least in terms of cost of ownership. The old-style analogue telephone network relied on the availability of large numbers of highly-skilled mechanical engineers and on a widespread electromechanical engineering technology pushed to its limits and now superseded. The electronic-based technology that has taken over has much more evolutionary capability and developing countries need to gain familiarity and expertise in that technology, not in an out-dated technology that is no longer cost-effective.

There is no simple binary differentiation between 'old' and 'new'. Development is a continuous process and we have to decide 'where to jump'. We do not have to make a once-and-for-all decision, however, concerning the totality of the network. It is quite feasible to envisage a broadband backbone network using ATM interconnecting residential networks using ISDN or digital PSTN technologies, with extensive radio coverage, enterprise network using X.25, frame relay etc., community networks using ISDN or B-ISDN video and image communications and so on.

The electronic long-distance highways face major financial problems, somewhat similar to those faced by roads and railways and, just as with these, require an immense infrastructure of standards and procedures before people can safely use the highway. Fortunately, the anarchy referred to above is less evident in the area of backbone network design and immense international collaboration has generated firm standards for the fibre-optic bearers, the electronics that provide the raw bandwidth and the synchronous digital hierarchy and the ATM layers that exploit that bandwidth to provide network- and transport-layer services.

Broadband infrastructure is being installed where it can provide present economies and, perhaps more importantly, future-proofing. Broadband ATM-based interconnect will become a widespread reality in Europe and North America within the next few years. It will be used to carry high-volume multiplexed telephony and low-speed data, as well as image and video services where these prove attractive.

These highways can ideally use terrestrial fibre-optic cable, (but fixed-channel radio and satellite links are perfectly valid components) and the cables networks themselves have a layered structure - they rely on protected ducts and providing the duct is often more costly than providing the electronics, so that there is a major incentive to equip telecommunications ducts along with roads, railways, canals etc., as a matter of course and so 'provide for the future'. The installation of 'dark fibre' provides another infrastructure capability that can be used by succeeding generations of optical systems, so that the end-to-end bandwidth available can grow from then on by equipping terminal and repeater electronics at specific points without touching the cable or the duct.

CONCLUSIONS

❑   We can see from the above that we should try to exploit all the 'good cards' at our disposal and beware of quick short-cut solutions. The concepts now being refined in terms of networking structures, although often based around SDH/ATM, are generally applicable and we should use the 'best thinking' that we can acquire.

❑   For backbone networks, fibre-optic cables have now reached a high level of maturity and there is little incentive left to use paired-cable or coxial in the long-distance network. SDH and ATM are rapidly achieving maturity and we should forget the idea that they can only be justified for broadband subscriber services: they will be used extensively to carry multiplexed low-bandwidth traffic. They will, however, provide a 'future proof' environment for further development, particularly for broadband image and video/multimedia services.

❏   Every opportunity should be taken to cut civil works costs by laying in telecommunications wired infrastructure (even if it is just the duct) whenever we can. Radio is a fine vehicle for telecomms but it is best seen as an adjunct to cabled networks, not as a universal solution.

❏   The general conclusion therefore is that advanced technology can and should be used in developing countries, but selectively, after a careful analysis of needs and opportunities, and with a full appreciation of its advantages and risks.

Telecommunications and Development in Africa
B.A. Kiplagat and M.C.M. Werner, Eds.
IOS Press

CHAPTER XVII

TRENDS IN RURAL TELECOMMUNICATIONS TECHNOLOGIES

Christoff Pauw
Programme Manager Telecommunications Systems
CSIR
South Africa

Telecommunications in South Africa has an exciting future. Although we have about the right number of telephones per capita in comparison with international statistics, I strongly believe that our economy can support a doubling of our telephone network over the next decade. India doubled its network from 1980 to 1990 [1,2]. Taiwan and Brazil achieved even higher growth rates in the seventies [3]. I believe South Africa can do the same from 1994 to 2004.

Raising the necessary capital will not be easy. I nevertheless believe we can and should set ourselves a national goal and then find a way to achieve it [4].

Technology is only one of the requirements for a good telecommunications infrastructure. Some other issues will therefore also be addressed in this paper. The emphasis is on Plain Old Telephone Service, because one must be able to walk before one can start running.

THE IMPORTANCE OF TELEPHONES AS A DEVELOPMENT PRIORITY

We all know that we cannot get our work done without access to a telephone. But what about rural villagers or inhabitants of urban squatter camps? The anecdotal evidence of the benefits of a community telephone in such a situation abound, eg [1]. Small-scale empirical studies have also been done [5].

Studies by economists prove the value of the availability of telephones beyond doubt. Kayani [6] quotes a study by Hardy which shows on theoretical grounds that increasing the telephone penetration from 1 per 100 population to 2 per 100 over a 10 year period, would directly cause an increase in GDP of 3,5%.

Some interesting figures emanate from a study in Botswana [7]. A rural expansion programme that would provide 5000 lines to 200 villages at a cost of US$7000 per line would have a total annualized life cycle cost of $1700 per line. Income generated would be $1200 per line per annum, requiring a subsidy of $500 per line. Now for the important part: the estimated benefit to cost ratio of a call from a payphone is a massive 8,8:1. And to the country's economy as a whole the *Economic* Internal Rate of Return of the project is a substantial 20%.

With this in mind, I would argue that the provision of community telephones to all communities where the penetration is less than 1 per 100 population should be the second highest development priority after the provision of clean water. I arrive at this conclusion very simply by looking at costs and benefits. In South Africa today there are very few, if any, villages where it would cost more than US$30 000 to provide a community telephone. One could drill a borehole and equip it with a pump for less than that, but what other service can you provide for US$30 000? How will you run the other services without communications? And keep in mind that a community telephone generates substantial income in all but the very poorest of communities, as was shown in [5], [7] and many other studies.

TELECOMMUNICATIONS POLICY AND REGULATORY FRAMEWORK

During the last decade major changes have swept the telecommunications services industry in the developed world. The industry has been liberalized and opened up to competition to a degree that was unthinkable two decades ago when most people still believed that the supply of telephone services would remain a natural monopoly for ever. The competitive forces thus unleashed have led to major strides in technological progress and to reduced costs.

In developing countries the trend is away from operating the telephone network as a government department. In a long list of countries, including Nigeria, Jordan, Sri

Lanka, India, Argentina and Chile the telecommunications operator was changed from a government department to a company owned by government, by government jointly with the private sector or even mostly by the private sector [3].

In my opinion the reason for these changes is that the provision of telecommunications services is different from most other services supplied by government in the sense that users pay for it, usually at rates that are not too far removed from actual cost. Therefore a telecommunications operator can obtain much improved efficiencies by carefully managing things like quality of service, customer satisfaction, cash flow, debt ratios and return on investment. This is quite different from other government functions such as policing and defence.

I firmly believe that without a proper policy and regulatory framework, and without clear separation of regulatory and operational functions, *unresponsive structures in the telecommunications sector will continue to hinder economic and social development and international competitiveness.* (Quoted from [3]. Also see [8] and [9].) The best technology in the world cannot fix problems caused by lack of policy and an in-appropriate regulatory framework.

Having said all that, let us now have a look at some exciting technological developments. Most of these technologies are available from South African suppliers. I will be mentioning costs because I personally feel that I cannot place technology in perspective if I do not have an idea of its cost. Naturally, readers are urged to approach the suppliers for more accurate cost information.

SELF-TARIFFING CARD PHONES

The humble payphone is a very important piece of technology in underdeveloped rural areas. Each payphone brings telephone service to hundreds of people who would ordinarily have no access to a telephone. And, in the Botswana study at least, lines connected to payphones generate the second highest income for the operator, after lines connected to village institutions such as clinics, schools, etc.

Traditional coin-operated payphones suffer from two important drawbacks in rural areas. These are the cost involved in collecting the money and the fact that the money in the box is too much of a temptation for thieves.

The card phones which are now being introduced into Telkom's network is a solution to this problem. The chip-cards are very safe, ie not prone to fraud, and cheap enough to sell in denominations of about US$3.00. The sale of the cards can be undertaken through small businesses, post offices, etc. The burden of collecting small amounts of money from many (remote) call boxes is removed from the operator and shifted to channels that exist anyway and that costs the operator nothing except a small commission paid to the sellers of cards. This kind of activity also fosters the development of small entrepreneurs, bringing much needed economic activity to rural areas [5]. One could even foresee such small entrepreneurs doing bulk breaking of the cards - selling a card for the balance of its credit plus a small handling fee and buying back the remainder after a call has been made.

In the South African network metering pulses are sent from the exchange. If required by other operators, the phones can be made to be self-tariffing in order to be able to connect them to any line without first having to install equipment for metering pulses at the exchange. Tariff tables could be downloaded via modem from a central control facility.

LOW COST SHORT HAUL MICROWAVE LINKS

Microwave technology made rapid progress in the last number of years. Microwave links offering a capacity of 2 Mb/s over distances up to 25 km are now available at a cost of below US$30 000 from several suppliers, and do not require expensive infrastructure to install. They typically operate above 20 GHz, where spectral congestion is not yet a problem.

Since collecting enough lines to adequately load a telephone exchange is a problem peculiar to developing countries, these links have a role to play to collect traffic from small neighbouring towns to a centrally located exchange.

LOW-CAPACITY POINT-TO-POINT LOCAL LOOP RADIO SYSTEMS

A radio link acting effectively as a wireless extension over tens of kilometres for two telephone lines are available at below US$5000. A range of more than 100 km can be obtained with the use of a repeater. A standard telephone service, including metering pulses, can be offered at the remote end.

Such links will find application to serve isolated users that can not be reached by other more economical means. They are also useful for temporary applications. For instance, a number were ordered for use during South Africa's April 1994 elections.

POINT-TO-MULTIPOINT RURAL RADIO SYSTEMS

The local loop is more problematical in developing countries than in developed countries. One dimension of the problem is in low-density applications where sheer distance is the main problem to overcome. Another dimension is urban systems where the local loop seems to be more difficult to install and to maintain than in first-world cities. Poor training of installation and maintenance staff is a problem, as is theft of copper cable. Wireless local loop systems are an attractive alternative, even if it costs more than conventional cabling.

Point to multipoint TDMA radio systems operating in the frequency band between 1.5 GHz and 2.5 GHz is a very important part of the arsenal of technologies available to operators in rural areas [13,14]. These systems are quite costly, generally working out at more than US$5000 per line, but they remain a good choice for low density applications, offering typically 300 lines over an area that may have a radius of 300 km or more.

The cost of systems in this category is likely to come down rapidly in the next number of years. The reason is the advent of wireless applications such as cellular telephones and cordless PABX's. The two standards DCS-1800, which is basically GSM at 1.8 GHz, and DECT, the Digital European Cordless Telephone, have spawned major technological advances and cost reductions. For example, the cost of a microwave filter in these bands came down from several hundred dollars to below $10. These technologies will find their way into rural systems in the very near future.

Many other technologies are available. Wireless local loop systems are available at around US$1,500 per line. This is total installed cost, including the base station and the outstation. Considering the problems with the local loop mentioned above, and the fact that the maintenance of wireless local loop systems is generally much cheaper than that of conventional cabling, this is an attractive alternative in urban and peri-urban areas.

Cellular systems also have a role to play. While it is probably not worth while putting up a cellular system just for rural applications, adding static rural lines to a system installed for other reasons makes a lot of sense. Part of the licence conditions of the two GSM operators in South Africa is that they should do just that. The necessary equipment is under development now (March 1994) and will be available soon.

Cordless telephones have a role to play in higher density areas such as central business districts and squatter areas. A DECT handset costs below US$300, and the base station is not very expensive either. Trials are being funded in Hungary by the EC to evaluate DECT as an alternative for the last 200 m of the local loop.

CSIR is developing a point to multipoint rural radio system relying on a new access technique called CDMA (Code Division Multiple Access). The system will have a range of about 30 km, relying on two outstation models, the one very cheap for close-in subscribers and the other more expensive and aimed at subscribers further away from the base station. The low-cost version is expected to cost below US$300, offering a range of up to 10 km. The system is expected to become available by 1997.

OPTICAL FIBRE

Optical fibres specially made for rural applications are available offering more bandwidth than coaxial cables at not much higher cost. Also, it has the all-important immunity against lightning damage, that plague of metallic cables.

One South African supplier has a cable featuring a central tube that can be fitted with between two and twelve multi mode or single mode fibres according to client needs. These cables can be installed on existing pole routes at below US$7,000 per km.

Androuchko [10] shows that upgrading an existing analogue two-wire trunk between exchanges (re-using the installed copper!) to digital technology is not worth while - if one has to upgrade it would be cheaper by a factor of five or more to install a fibre route instead.

The low cost of fibre is also spawning all sorts of local loop applications. A distributed concentrator system developed for rural applications is described in [11]. Kinnear [12] has done some comparative cost studies showing that this system is substantially cheaper than rural radio systems in some applications.

It is widely believed that fibre in the local loop will be cheaper than copper before the turn of the century.

FIBRELESS OPTICAL SYSTEMS

Point-to-point free-space optical systems are not widely available as commercial systems, but are known to exist [15]. Such systems may have a range of up to 15 km and offer data rates over 1 Mb/s at a lower cost than short-range microwave systems. Also, no radio spectrum and no frequency management is required. Free-space optical products should become available in the next year or so.

A very interesting development is that of an optical point to multipoint local loop telephone system now underdevelopment at CSIR. Subscribers can be served in a range of 400 m around a base station. The projected cost is just $20 per subscriber station. It is aimed at high-density settlements such as the squatter areas found around many cities in developing countries. The equipment is meant to be self-installed by the subscriber. The subscriber can also be asked to bring the equipment to a maintenance facility in case of malfunction, thus freeing the telephone operator from the expensive task of maintaining the local loop.

The product is expected to be available by about 1997. A demonstration system is available for viewing at CSIR in Pretoria. Telecommunications manufacturers interested in commercialising the technology may approach CSIR.

SATELLITE SYSTEMS

Satellites have a role to play, especially to provide service rapidly to sites that cannot easily be reached by radio. Businesses requiring high-speed data channels also often find a small satellite terminal the best short-term solution, at least until terrestrial systems can reach their site.

For travellers and contractors working at remote sites Inmarsat is a real blessing. Africa is a major market for Inmarsat. Costs are fairly high, but affordable for business users. Brief-case sized Inmarsat data terminals are available for below US$10,000. Operating costs are quite affordable for short typed messages, but becomes prohibitive if large amounts of data need to be transmitted.

Both Intelsat and Panamsat will soon be offering transponders suitable for use with cheap VSATs (Very Small Aperture Terminals). Panamsat's PAS-3 satellite is scheduled for launch in 1995 and will offer C-band coverage of the entire African continent at 38 dBW. The Republic of South Africa will also receive a Ku-band spot-beam at 52 dBW, which is suitable for a 64 kb/s earth station that costs below US$10,000, employing a 1.2 m dish. Transponder rental is said to be very reasonable. Unfortunately, most administrations use satellite communications to cross-subsidise other activities, thus making it very expensive for the end user.

I am often asked in connection with the rural telephone systems now under development at CSIR whether such systems will not be made obsolete by low earth orbit satellites such as the Iridium system. The answer is an emphatic no. Iridium will only support a few million subscribers (presumably at a low call density) and calls will cost US$3 a minute [16], which is way beyond what can be afforded by a typical African villager, and has to be paid in foreign currency as well.

Satellites remain one of the more expensive solutions that are mostly used when terrestrial solutions are for some or other reason too slow, difficult or expensive to implement.

ADVANCED DIGITAL SERVICES

I said in the beginning that the emphasis would be on telephone service. What about digital services? My belief is that in rural areas operators should not invest large sums of their own capital in infrastructure specifically aimed at advanced digital services. Data services should be supplied on a demand-led basis, unlike basic telephony which should be supplied on a supply-led basis.

Data services are usually required by large companies. For example, if a mining operation is to be started up in a remote rural region, then the mining company should assist the telecommunications operator to find the financing to supply the needed data services. Alternatively, if the operator can not offer added value to the mining company's efforts, such a company should be allowed to provide for its own needs, perhaps via a private satellite earth station. The telecommunications operator should not hinder economic development to improve its own fortunes.

Having said that, I think that it is very important to use digital technology wherever possible when networks are expanded or old equipment replaced. Certainly, exchanges other than fully digital are hardly being manufactured any more. Trunk routes and concentrator feeds should also be digital. Even in wireless local loop applications, I think that one should be wary of investing in analogue technology, unless there are substantial cost advantages.

COOPERATION BETWEEN AFRICAN COUNTRIES

In telecommunications volume is one of the most important determinants of cost. It applies to the purchasing of equipment, to training, to managing a network, to maintenance, to almost every aspect you can think of. Since all of Africa has less telephones than countries like France or Germany, it would be to our great advantage if we could work together using common standards and equipment. If a number of countries could coordinate their expansion programmes with common suppliers, larger production runs would result in useful cost reductions.

SOUTH AFRICAN NETWORK OPERATORS

South Africa has five network operators, all of whom have strategies to expand their business into Africa in the shorter or the longer term.

Telkom SA is the dominant operator with 3,7 million main exchange lines in operation. It operates international circuits to 53 countries. It is a share-holder in the SAT-2 optical fibre cable linking the Southern tip of Africa to Europe. This cable is currently the cheapest transport medium for traffic from Southern Africa to Europe. One would hope that the present process of political change will enable South Africa's neighbours to benefit from SAT-2.

Eskom is the national electricity supplier. It has its own telecommunications network and cooperates with several African countries. High voltage electric cable routes crisscrossing Southern Africa are ideally suited to carry fibre optic communications cables as well.

Transtel is the telecommunications subsidiary of Transnet, South Africa's government-owned railways and transport company. Its railway carriages and locomotives operate far north of South Africa's borders. It has a network of offices and agencies, including South African Airways offices, in many African countries. It is currently expanding its capabilities to be able to link all its African operations via satellite. This capability can reasonably be expected to be available to other business entities as well.

South Africa has two competing cellular telephone operators, MTN and Vodacom. Both are using the GSM standard. Their licences were awarded late in 1993 and they are at the time of writing in the throes of getting their networks fully operational by June 1994.

Many of the technologies described above are available from South African equipment manufacturers, some of whom are subsidiaries of multinational companies. Among these suppliers are: Aberdare Cables, Alcatel Altech Telecomms, ATC, Plessey

Tellumat, Siemens Telecommunications and Telephone Manufacturers of South Africa.

CSIR - RESEARCH PARTNER

CSIR is a statutory council with operational autonomy within a framework set by the South African government. It has a board appointed by the Minister of Trade and Industry consisting of highly respected individuals from business and the community at large. Its budget is made up of a parliamentary grant (45%) and contract income (55%). CSIR has about 3000 employees.

CSIR'S MISSION IS TO:

❏   be the technology partner of South African industry in both the formal and informal sectors to promote economic growth;

❏   provide technology solutions that improve the quality of life in urban and rural developing communities;

❏   provide scientific and technological support to enhance decision-making in the public and private sectors.

Although South African industry is mentioned in the mission statement, CSIR works for clients world wide, including clients in 13 African countries. It is the largest R&D organisation in Africa. Its activities range from forestry to building and transport technology, materials science, mining technology and aeronautical systems.

CSIR's Division for Microelectronics and Communications Technology has 270 employees, mostly engineers, scientists and technologists. Its activities include work in the fields of radio systems planning, computer networks, secure communications, electronic identification, geomagnetic surveys, magnetometers, spacecraft tracking and telemetry and earth resource imagery. Research and development services and consultation is available in wide range of disciples. Investors are also sought to participate in the exploitation of intellectual property.

OoOoOoO

REFERENCES

[1]   Pitroda, S: "Development, Democracy and the Village Telephone," *Harvard Business Review*, Nov-Dec 1993.

[2]   Siemens: 1992 International Telecom Statistics.

[3]   Wellenius, B: "Beginnings of Sector Reform in the Developing World", *Restructuring and Managing in the Telecommunications Sector*, The World Bank, Washington DC, 1987.

[4]   Pauw, C K: "Private enterprise in the local loop and the road to universal service,"   Conference on Telecommunications for Rural and Township Communities, Johannesburg, 15-16 March 1994.

[5]   Morris M L and Stavrou S E: "Rural Telecommunications", *TELKOM 91 Proceedings*, Sun City, Nov 1991.

[6]   Kayani, R A: "Impact of telecommunications on economic development", Tanzania Posts and Telecommunications Corporation, ca 1984. Obtainable from mr Kayani, who is now at the World Bank in Washington DC, or from the author.

[7]   Clarkstone, A D *et al*: "Rural telecommunications in Botswana: socio-economic and strategic issues," *Proceedings of the Second International Conference on Rural Telecommunications*, IEE, London, 1990.

[8]   Kayani, R A: "Rural telecommunications development policy in developing countries," *Proceedings of the International Conference on Rural Telecommunications*, IEE, London, 1988.

[9]   Kayani, RA: "The impact of liberalization and sector restructuring on rural telecommunications development in developing countries," *Proceedings of the Second International Conference on Rural Telecommunications*, IEE, London, 1990.

[10] Androuchko, L *et al*:   "Optical fibre technology for developing countries," *Proceedings of the Second International Conference on Rural Telecommunications*, IEE, London, 1990.

[11] Levieux, C: "The use of optical fibre to provide economic telecommunications in rural areas," *Proceedings of the 1993 IEEE South African Symposium on Communications and Signal Processing*, IEEE Catalog Number 93TH0546-2, New York, 1993.

[12] Kinnear, K: Unpublished tutorial, IEEE Africon '92, Swaziland.

[13]   Morris, M J: "Planning and implementing a rural telecommunication system using microwave TDMA subscriber radio," *Proceedings of the Second International Conference on Rural Telecommunications,* IEE, London, 1990.

[14]   Culot, B *et al*: "Rural Communication: the global concept," *Proceedings of the Second International Conference on Rural Telecommunications,* IEE, London, 1990.

[15]   Chaimowicz, J C A and Cole, R S: "Fibreless optical communications links for isolated communities," *Proceedings of the Second International Conference on Rural Telecommunications,* IEE, London, 1990.

[16]   Mason, C F: "Iridium forges ahead with its grand PCN plan," *Telephony,* November 1, 1993.

Telecommunications and Development in Africa
B.A. Kiplagat and M.C.M. Werner, Eds.
IOS Press

CHAPTER XVIII

COST EFFECTIVE RURAL COMMUNICATIONS USING FIXED CELLULAR RADIO ACCESS

Rudi Westerveld
Telecommunications and Traffic-Control Systems Group
Delft University of Technology
The Netherlands

Telephone penetration in Sub-Saharan Africa has an average ratio of 3 telephone subscribers per 1000 inhabitants, mainly concentrated in the urban areas. The largest part of the population living in rural areas is deprived of telecommunication facilities. The overall growth of any country in this time depends a lot on its telecommunications facilities. Therefore, it is necessary that proper telecommunications systems should be developed that are also affordable for African rural zones. The telecommunication system should be planned such that it puts a minimum and affordable financial burden on the rural user. Considering this point, a fixed cellular radio system is proposed. This system should be combined whenever possible with an existing mobile network to reduce initial and operating costs. The fixed cellular system can use adapted readily available cellular mobile transceivers. Thus it may save the cost of developing and manufacturing of new radio equipment for the terminals.

Cellsizes can be increased to cover a radio range in the order of 50 kms because of the fixed position of the "outstation".

A fixed cellular radio system is not a completely new concept. Single user versions are already put in operation in several countries for rural communications. The concept of a fixed rural radio systems has been discussed in numerous papers and several companies supply these systems. This paper proposes an extension of the concept using multiple cellular radio links to provide trunked access for a small village exchange with as few as 8 subscribers. Also a comparison of these systems with the classical point-to-multipoint systems is presented.

The main motivation for writing this paper is to give a wider publicity to the application of fixed cellular systems for providing telecommunication facilities to rural people in developing countries.

This paper presents both economical and technical aspects.

BACKGROUND

Rural communication especially in developing countries has been a topic of discussion for quite some time. Different solutions have been presented. But none have really solved the problem, mainly because of the fact that the proposed solutions are still unaffordable to rural subscribers.

The need for communications in rural areas in developing countries has been agreed upon in many studies. Arguments in favor that were used are: National integration of remote areas, reduction of the need to travel, emergencies, catalyzer and tool for rural development, support for trade and tourism. Chowdary [8] sums up why access to rural telecommunications in the case of India is desirable and justified. Most of it is valid as well for rural Africa.

❏ for distribution of the benefits of science and technology to all sections of the Society, including people removed from urban habitations and activities.

❏ as an instrument enabling rural people to be drawn into economic activities and exchanges on an extended scale.

❏ for conferring beneficial marketing for the agricultural surpluses being brought about by improved, scientific and irrigated farming; and

❏ for reducing the isolation from the educated sons who had migrated to the cities, distant towns and lands.

Rural telephony must be looked upon as a development tool, and not as a commercial service. We have to induce demand as well as meet it innovatively and beneficially.

The main issue for telecommunications however is: *can rural areas in developing countries be cost effectively covered?*

The general opinion is that the provision of rural telecommunications is often a high cost, low revenue business. A small proportion of the total amount of subscribers can be a large burden on the backs of Public Telecommunication Operators (PTO). The high cost of "wiring and rewiring" service areas has been cited as a major reason for low penetration of telecommunication networks in rural areas (World Bank Statistics)[2]. Other reasons are: low telephone demand, low utilization, dispersed subscribers, difficult terrain and remoteness from the national network, low generated revenues[5].

One of the reasons for low revenues is the lack of proper maintenance and hence a low grade or even a complete lack of service. The poor availability of the systems is a logical consequence of neglecting essential maintenance. Initial subscribers, often business and government, that have a real interest in using telecom services for their business, are turned away and will be looking for other ways to solve their needs. Through a lack of demand as a consequence of providing a bad service at a high cost, the necessary revenues will not be generated and services can not be provided in a profitable way.

Radio based technology has been developed to overcome some of the mentioned problems. This paper discusses the use of advanced radio technologies, specifically cellular radio systems for fixed rural subscribers. A comparison has been made with the existing rural radio systems. Also the migration from analog systems to digital systems is discussed.

COST OF RURAL SUBSCRIBER ACCESS

GENERAL REMARKS

In discussions about the cost of providing telecommunications facilities to rural areas confusion exists about who is paying what. We can first of all analyze the value chain for the provisioning of rural telecommunications. This chain starts at the manufacturing plant, then the operator comes into play and finally it is the rural subscriber who really has to pay the telephone bill.

The price a manufacturer covers for the equipment has several elements: a pure equipment manufacturing cost element, a quantity element and also a lot of commercial and marketing oriented aspects. It is very difficult to get price information from manufacturers as they will respond with the following questions: How many lines do you need, for which country, what project etc.?

Of especial interest in planning a network is not only the subscriber density, as is often used in models of rural areas, but also the so called grouping factor for rural subscribers; how many subscribers can be served by one access point.

Once a buying decision is made the next step is to install and commission the equipment. Most of the work is done by the operator; but the manufacturer will be involved also. A large component of the expenses of the operator is spent on providing transport, housing and/or shelter for the equipment.

And after having installed the equipment, the operation and maintenance cost have to be considered. This part of the value chain is very much dependent on the local situation and can vary considerably. Finally we come to the rural subscriber and the amount of money he has to pay for the telephone service. That depends on many different factors related to local government and operators policy; e.g. cross subsidization policy, loan or grants conditions by international institutions, the way investments are accounted for, rate of return on investment used and depreciation rate, existing infrastructures etc. And last but not least the foreign currency situation of a country plays an important role as most of the equipment has to be imported and paid for in a convertible currency.

Exact data are difficult to be found. But to get an idea of the relative values we can look at the revenue chain for general telecommunications procurement in Europe see figure 1 (source System Dynamics).

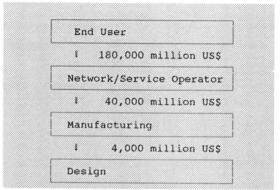

Figure 1 Revenue chain of Telecom procurement in Europe

Here we see that the major amount of money flows from the end user to the service providers. Although the situation in other continents may be different,yet it gives an indication of where the focus must be given in discussing possibilities of providing rural telecommunications. Manufacturers prices only give a bleak indication of what the real costs are.

Concluding one can say that in discussions about cost of a rural telephone line per subscriber, we have to establish clear conditions and to agree on realistic models. Only then we can make meaningfull comparisons between different solutions. In this paper however we keep on using the "cost to operator" as we have no other model sofar.

COMPARISON OF TECHNOLOGIES

EVOLUTION OF SYSTEMS

In the design of rural telecommunication systems the main consideration is for: cost, noise, coverage, switch capacity, reliability and transmission cost [1].
There has been a tendency to overemphasize the role of switching with rural telecommunications. But in the rural areas, transmission is even more important, especially if direct dialling is to be given and reliability and satisfactory quality are to be ensured. There is even greater need for research and development of transmission systems appropriate to rural areas [8].

For many years radio systems have played an important role to provide telephone services to subscribers at remote locations. The earlier systems were using VHF\UHF narrowband FM single channel radio.
Later on, some 10 years ago, point to multi-point systems using FDM and TDMA were successfully introduced. They provided a major improvement compared to the older systems and gave an almost transparent connection into the national and international network.
Recently a new class of digital TDMA radio systems has come to the market (main suppliers are TRT, NEC, SRT and Alcatel) with an improving performance-to-cost ratio [5, 11]. These systems can not only provide high quality voice and data telecommunication services but will be able to provide interconnection to ISDN.
The available systems can provide services to a wide range of small to large subscriber groups. Low capacity systems have 10 to 15 trunks and have served groupings of approximately 100 subscribers. Higher capacity systems have become available with 30 to 60 trunks that can provide service to approximately 1000 subscribers.

RADIO TECHNOLOGY

Radio technology has been applied in rural telecommunications for some time now. Analog Point-to-Multipoint (PMP) systems are provided on the market by most manufacturers of telecommunication equipment. Cost per subscribers has come down to approximately $ 2000. New markets have been opened in Eastern Europe where PMP systems are now used not only in remote areas for local access, but also in semi-urban areas as the so called radio in the local loop. The suitability of these technologies lays in the speed at which local radio distribution can be provided. This is likely perceived to offer a significant advantage. Radio systems can compete with conventional copper wire links on the following grounds: initial cost, installation time, flexibility, adapatabilty to terrain, reusability of equipment and maintenance cost.

SATELLITE TECHNOLOGY

Satellite voice communication has been for a long time far out of reach for the non-institutional rural subscriber in developing countries. The size and cost of antenna and radio equipment were not affordable for small rural communities. So only in emergency and natural disaster situations transportable terminal equipment was brought in, to access a satellite telecommunication service (mainly through geostationary Inmarsat-A).

The provision of direct access voice services is developing rapidly in various directions. There are several options to choose from for a single user who wants a point to point voice service from a remote rural area . These services can be accessed now through briefcase size portable terminals and in the near future even handheld terminals will become available. These handsets will provide access in a similar way as cellular systems that are used in urban areas.

Two approaches are emerging, the first is the use of large geostationary satellites (altitude 35,780 km) with large satellite antennas and high frequencies (20 to 30 GHz) resulting in highly focussed beams known as spot beams.

In the second approach low-earth orbiting (LEO) satellites (altitude 780 km) are used, with the reduction of the path loss as the main advantage. The LEO approach also gives much smaller propagation delays than those experienced with geostationary satellites (single hop maximally 12 msec versus 0.25 sec in the geostationary case).

In both approaches there is another significant development towards smaller satellites (microsats) and smaller terminals.

There is an enormous potential for mobile and remote rural services for microsat communicatons constellations. Both on a performance and economic basis they have a potential to compete with terrestrial systems. They provide reliability through redundancy, cost advantages through economies of scale and gracefull introduction of capacity and new technology upgrades. LEO constellations use lower frequency bands resulting in the possibity of use of less expensive omni directional subscriber antennas.

LEO microsat constellations when properly integrated with other terestrial and space systems can provide a major advance in opportunities for communication with and within rural areas in developing countries.

Among the many new enabling technologies microelectronics, digitization, high-speed on-board processors,new signalling and switching devices, more efficient signal coding, voice and video compression, advanced satellite antenna design and ground terminals like VSATs and USATs played a crucial role in the rapid development of cost-effective microsat systems. An indication of the increasing interest in the deployment of systems based on LEO microsats is the amount (over 280) of FCC filings for satellite services on LEO satellites. Among them are Iridium with 77 microsats, Loral Cellular Systems Globalstar with 48 microsats and Italspazio's LEOCOM with 24 microsats.

COMPARING WIRE AND RADIO

The use of wirelines in rural telecommunications has been considered as a barrier for rural telecom growth [4]. Some arguments are:

❖   Worldwide average cost of copper service $2000/subsc. (World Bank statistics) is high and stable.
❖   Wirelines pose physical and geographical limits.
❖   Wirelines are less flexible and create difficulties in long term planning.

A price comparison for an U.S. urban area [2] indicates that over the time span 1990-1996 the wire cost will stay stable at approximately 800 US$/subscriber and the cost per subscriber of cellular radio local loop provision will go down in that period from 1000 to 700 US$. Another problem with wirelines lies in the fact that there is a high ratio (10:1) between the highest and lowest cost of wirelines in a rural network [14].

With radio based technologies the initial investments are lower. Only the handset is a fixed cost per subscriber. The network can be dimensioned towards "day 1 traffic", further growing can be paced according to demand. This is important for developing countries where funds are particularly limited. Another option is in initial limited fixed (stationary) voice service, with later extension to data and mobile service. In this way infrastructure investments can be phased [3]. Ratio of highest cost to lowest cost is much more favourable (2:1) with radio. Base stations and terminals in radio systems usually have a substantial recovery value, unlike copper plant. The salvage value of the latter is about equal to the cost of removing it [14].
This can be very important in networks where demand is not very well predictable and reallocations and changes have to be made frequently.

Many manufacturers are offering now solutions with relatively cheap and small subscriber terminals for internal desk or wall mounting. These systems are not only offered for use in the developing countries but are specifically aimed at rapid and flexible provision of telephone services in Central Europe. This makes it even more attractive for use in developing countries because prices will go down when higher production volumes emerge.
Another favourable aspect for radio systems is maintenance cost. Radio systems are more reliable and radio base stations can be positioned at relatively safe places (e.g. government buildings) to prevent tampering. Maintenance and fault diagnosis can be remotely done through more sophisticated systems [3, 10]. Personel with higher skills are needed in that case, but with less personel than required for wireline a larger area can be covered.

FIXED CELLULAR

However cost is probably still to high for rural areas in developing countries. Therefore another solution has been proposed and implemented [1], [5] and [6], namely the use of existing cellular mobile telecommunication systems (CMTS) with fixed rural subscribers (CMTS-F). Rural subscribers in the vicinity (distance < 50km) of large urban areas and roads could be connected to a CMTS and use the excess capacity of that system. Main advantage is the sharing of cost of installation, operation and maintenance. In this way the relatively rich mobile subscriber will subsidize the poor rural subscriber. Also there will be the combined use of the already to the mobile system allocated frequency spectrum. Initial users will be police, hospitals, ambulance and other emergency services, gas stations and Public Call Offices (PCO) or coin box telephones.

SMALL SYSTEMS

All these systems have in common that they can provide rural service in a cost effective way only to relatively large groups of subscribers.
Especially for small villages in developing countries there is still an initial need to provide service to a very limited amount of subscribers. When we try to add up the most likely first subscribers we do not get more than about 4 (e.g. a Public Call Office (PCO) or other type of community telephone, a Government/Police post, a health care center and one merchant). Some of the modern systems can hardly provide cost effective service to such small groups of subscribers. Therefore new radio systems are emerging that can connect single subscribers in a cost effective way. These systems rely on self-contained, small and low cost radio terminals at the subscribers premises. Some systems are a further development of the "classical" PMP systems, such as the TRT IRT2000 microstations. Others make use of cellular radio systems originally designed for mobile subscribers. Table 1 summarizes the "single subscriber" systems with their main characteristics.

TERMINALS

The terminal that is used in these systems is sometimes contained in a small flat box, with a short whip antenna that can be placed under and connected to a standard telephone set. Others provide a wall or pole mounted box with an attached antenna and a normal connection for a telephone set. The systems that work in a mobile cellular network mostly use a commercial mobile transceiver without a handset and a battery. They are equipped with an adapter unit to provide an interface to an ordinary telephone set, fax machine or modem. These adapters provide for 4-wire to 2-wire conversion, support for ground or loop start, pulse or tone detection, ringing current and power supply with internal back-up in case of power interruptions.
Some of the systems are equipped with a more sophisticated interface unit for connection to a PCO. They have to provide tariffing signalling, answer detection and different types of other signalling to the pay phone. Sometimes an internal modem is added so that the stored tariffing information can be modified remotely from an O&M center.

COMPARING DIFFERENT RADIO TECHNOLOGIES

Diaz-Hernández [9] reported an interesting comparative analysis of different radio technologies for rural communications.

Manufacturer, type	System	Freq MHz	Terminal	Max. Distance	PCO interface
NEC, DRMASS	PMP	1500–2600	Wall/pole mount	<45 km	yes
TRT, IRT2000, Microstation	PMP	1500–2500	Wall mount	?	?
Ericsson, RAS1000	PMP	380–500 or 800–1000	Table model	<80 km	no
Tadicom, TADiCell	PMP	824–878	Wall mount	<10 km (<20 km)	no
InterDigital, UltraPhone 100i	PMP	450–460	Wall/pole mount	<60 km	yes
Ericsson, ATUR	NMT, cellular	450–470	Wall/pole mount	*)	yes
Nokia, WLL	NMT, cellular	450–470, 900–960	Table	*)	no
Codecom, TEL–CEL 200	AMPS, NMT, ETACS, cellular	820–890, 450–470, 870–950	Wall mount	*)	yes
Novatel, SCS–Libra, suppl. cellular syst.	AMPS, cellular	820–890	Mobile	*)	no
Siemens, GSM PCO	GSM	905–960	Wall mount	*)	yes
JRC, SUD 401	Proprietary cellular	400	Table, Mobile	30–50 km	no

*) Depending on designed cellular coverage.

Tableau 1 Single subscriber radio systems

A comparison was made between a Digital Multiple Access radio Point-to-Multipoint (PMP) system, a Cellular Mobile GSM system with fixed subscribers (GSM-F) and a low earth orbit multi-satellite (48) system GLOBALSTAR (STAR).

Their impact was analyzed for three scenarios: years 1992-3, 1992-7 and 2000. The study was done with an application example in a typical rural area focussed on the access network segment of the GSM-F, Star and PMP technologies.
Uniform subscriber distribution was assumed, with subscriber densities from 1 subscriber per 10000 square kilometres to 20 subscribers per square kilometres. The

following subscriber grouping factors were used: 1, 8, 16, 50 and 100 subscribers per point.

The 1992-3 and 1996-7 scenarios are both a comparative analysis of GSM-F and PMP technologies. In the latter scenario the cost of equipment is reduced by 30%, and the cost of line transmission and infrastructure is considered to be the unchanged. The scenario 2000 includes the STAR technology which will probably enter to provide service around the year 2000.

The results of the analysis for the first two scenarios indicate that GSM is favourable considering cost for subscriber densities higher than 0.02 S/km^2 and with a grouping factor less than 16. The PMP system can provide cost advantage with subscriber density from 0.0001 to 0.02 S/km^2, with any grouping factor, and from 0.02 to 20 S/km^2 with a grouping factor of more than about 16. There is a small group of subscribers: very low density (< 0.0002 S/km^2), but high grouping (> 50) where GSM is still favourable.

The scenario for the year 2000 positions STAR between GSM and PMP. STAR will provide cost advantage for densities lower than 0.02 s/km^2 and grouping lower than approximately 16. Figure 2 gives a summary of these findings.

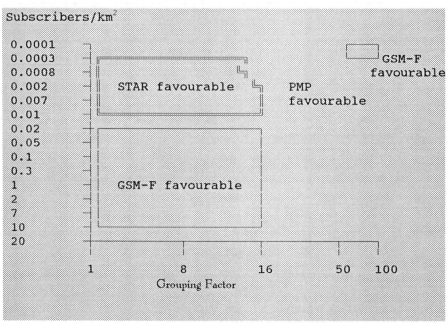

Figure 2 Preferred areas of different systems

According to this study [9] the trend will be that for rural areas with a low grouping factor GSM-F and STAR technologies will grow and PMP will decrease for next generations. The PMP technology will continue to offer the most cost-effective services within rural areas that have moderate to higher grouping factors (>16).

FIXED CELLULAR

CELLULAR MOBILE BASICS

The basic design ideas for cellular radio telephone systems are: high capacity, large number of subscribers, relatively low powered base stations, reuse of frequencies, resulting in small cells(>10km radius) [1].
Cellular radio gives: repeated use of frequency, increase in system capacity and channel assignment on demand [4]. It is a means of providing high density communications without consuming large amounts of spectrum (frequency reuse) [2].
Competition is now taking place between ADC American Digital Cellular versus GSM Groupe Speciale Mobile. Which system will give more increase in capacity through the use of digital techniques (Time Division Multiple Access TDMA or Code Division Multiple Access CDMA) [2]?

COMPARISON BETWEEN FIXED AND MOBILE CELLULAR

Advantages of Fixed cellular systems to mobile [4]:

❖ simpler logic unit -> lower cost radio
❖ no handoff -> simpler and cheaper switching
❖ high antenna's permit line of sight links and less propagation loss.
❖ lack of motion no short-term multipath fading -> smaller S/I requirements.
❖ directional antennas give increased frequency reuse and capacity.

Cellular networks in rural ares for fixed subscribers can be provided with reduced cell site costs because of less equipment is required (no handoff and Raleigh fading) [1].

COVERAGE

Fixed cellular systems can have a maximum cell radius of > 60km. Because of the radio horizon effect, there will be less co-channel interference. This leads to a minimum co-channel separation (A normalised reuse distance D/R value of 3 might be sufficient) [4].

CAPACITY

Cellular mobile systems normally provide ample capacity. Average rural traffic (developed world) is estimated at 0.25 E/Sq.km. This traffic level can easily be provided by radio systems [3]. However when systems are designed for thin route coverage along highways, special care has to be taken in the capacity design to provide for the ability to mix different traffic patterns. Fixed subscribers can have a completely distinct traffic pattern from a mobile subscriber. This is for instance the case when a PCO in a village is connected. Traffic generated from a pay phone can be quite high (0.25 E).

DIGITAL VERSUS ANALOG RADIO SYSTEMS

Digital cellular is already in the early 1990's more advantageous than current analog types because of

1.   Cheaper than analog through extensive use of VLSI and

2.   More efficient spectrum use through TDMA (2 to 3 times) (American CDMA even more efficient?) [3].
      Digital cellular radio means not only voice but mostly also easy data. Even ISDN is possible [3].
      Digital radio provides reasonable inherent security from eavesdropping and unauthorised line use [3].

ADVANTAGES DIGITAL VERSUS ANALOGUE

❏   Digital:
     ❖ prices are going down through mass production
     ❖ Equipment is more compact through VLSI
     ❖ More efficient use of radio spectrum
     ❖ Higher transmission quality
     ❖ Data service possible without modem [3].

❏   Analogue:
     ❖ It is available now
     ❖ It will remain cheaper for next 2 to 3 years
     ❖ Maximum radius is larger because of lack of equalisation for transmission delays [3].

EXISTING SYSTEMS

SYSTEM DESCRIPTIONS

ATUR system (Automatic Telephone Using Radio) Malaysia [6]
Malaysia has pioneered the concept of using the mobile telephone system to provide telephone service to rural areas via Coin Collecting Box (CCB) public telephones. A special radio interface unit was developed by Ericsson for Malaysia which makes it possible to connect the standard CCB telephone to the ATUR network and thereby eliminate the physical wire connection. There are approximately 200 CCB's in operation.

The distance from the coinbox telephone to the radio unit can be 2 km. Antennas can thus be mounted on a nearby hill or high building. Power comes from the mains supply or from solar powered batteries. The transceiver is connected to a directional Yagi antenna, so that the unit can be placed even further away from the base station site than is possible for normal mobile telephones.

SAUDI ARABIA

In an article describing the telecom development in Saudi Arabia including the Automatic Mobile Telephone System a case study is mentioned for the installation of a combination of fixed rural subscribers and the mobile cellular system [5].

RURAL COMMUNICATIONS VILLAGE PROJECT

VILLAGE NETWORK

INITIAL CONDITIONS

In the near future we may not expect many domestic subscribers in African villages because of the very limited financial capacity of most villagers. Most initial subscribers will be found among government/police posts, small medical posts/hospitals, store owners and some payphone/PCO. Thus we expect an initial demand of approximately 4 to 8 subscribers per village. The design of a village telephone system has to be cost effective for such a small amount of subscribers. Therefore a concept is chosen that relies on modern but simple digital switching technology for the small "village exchange" [12] and for the use of fixed cellular technology for the connection to the outside world

SYSTEM CONCEPT

The system consists basically of a small local switching node to which max. 8 subscribers can be connected. The system is self-contained and does not need external control for the completion of local calls. Service to the outside world is provided by two trunk lines using fixed cellular connections through two radio transceivers providing access to an existing mobile cellular network. No changes or adaptations will be needed on the network side of this mobile system. All necessary adaptations can be implemented on the village system side.

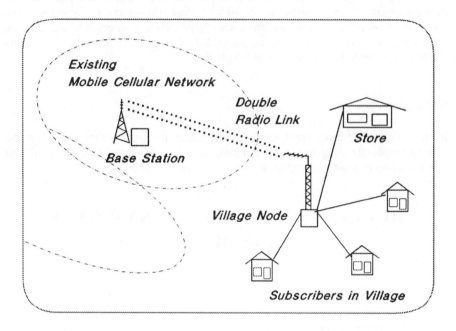

Figure 3 : Overview village telecommunication system

The two transceivers can share the antenna system. The use of two transceivers brings two main advantages: improved availability of the system and higher capacity. Of course there also is a disadvantage of an increase in cost. However especially the higher availability is of prime importance in a rural area in a developing country and worth the extra investment. In case of failure of one of the transceivers there will be a "graceful" degradation of service, without a complete isolation of the system. This also permits uninterrupted access for centralised maintenance and service restoration purposes.

CAPACITY AND GRADE OF SERVICE

For the design of a rural village telecom system one has to find a balance between cost, capacity and grade of service. To determine the grade of service that can be provided at a reasonable cost it is also necessary to have proper data on subscriber behaviour. These data are not readily available, but on the basis of research from India[13] we can say that with an estimated amount of traffic of 0.05 erlang per subscriber the system can accommodate 7 subscribers on two outgoing lines with a 5 % loss of traffic. Village traffic is not affected by external congestion as it is handled by the local switching node. The effects of congestion on the external lines can also be alleviated by features of the local switching node like automatic queuing of trunk calls, automatic call-back, emergency seizure of trunks and call timing.

When a PCO needs to be connected with a relatively higher traffic load to the village node, it might be advisable to provide it with a radio transceiver and interface set of its own. These systems are already available from different manufacturers.

TERMINAL DESIGN

Connection to the outside world is made through two radio transceivers that can connect to an existing mobile cellular radio system. The transceivers are modified local cellular standard transceivers. The modification enables the extraction of signals to provide inputs to the Interface unit. These are the 4-wire speech signals, dialling signals and hook signal. Also some modification of the transceiver control software is necessary to make the called number information and other signalling accessible during the call set-up process. The transceivers are connected to an interface unit that brings together the different signalling formats and protocols of the transceivers and the digital switching node. This node can provide access for a maximum of 8 subscribers. The call handling is completely autonomous at the village level.

There are two trunk lines implemented for connection to the interface unit. The node has many features to be found in modern exchanges likes automatic call-back, priority queuing, call timing and emergency break-in which are especially useful when congestion conditions exist at the outside lines. Last but not least there is a supervisory unit that takes care of all supervision and maintenance functions for the village station. It keeps contact with a central maintenance unit and preprocesses the relevant local information.

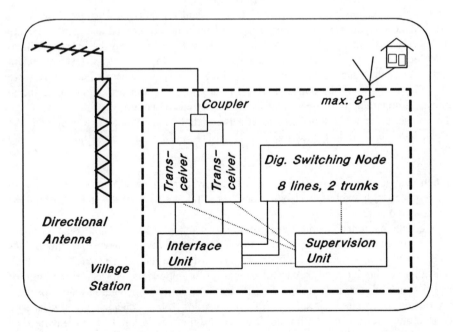

Figure 4 Village Terminal

In the design attention is paid to low power consumption so, depending on local power supply conditions, solar powering can be considered for this station.

MAINTENANCE

RURAL CONTEXT FOR MAINTENANCE

Providing a technical "solution" to the rural telecommunications problem for developing countries is one thing. To provide a satisfactory solution is something else. In many reports the importance of telecommunication for rural development is stressed. However little is said about the demotivating effect that results from the provision of a new telephone facility, that breaks down after a few months and is often only available for operation after many months.

Apart from the loss of revenues when a failure occurs, one has to consider also the indirect effects of a unreliable network. When people can't rely on a working telephone system, they can hardly be persuaded to use it and to pay for the subscription. They will not start any activities in which they become dependent on telecommunications and the often expected economical stimulus of the introduction of telecommunication services is nullified.

The rural environment, in many developing countries in particular, puts a lot of stress on equipment. There are not only the harsh climatological conditions that influence the availability of telecom equipment. Failure of power systems and unforeseen human interventions also cause interruptions of operation.

And when a system breaks down maintenance and repair is hard to get. Centers of maintenance are at great distances. Transport facilities are limited and rural areas are often very difficult to access in certain periods of the year. Also the limited skills of field technicians and the lack of spare parts, proper documentation, test equipment and other maintenance support impede rapid repair. Hence the Mean Time To Restore (MTTR) or Time of non-availability (Ti) in developing countries is quite long and in no way comparable to those in the western industrialized countries.

To calculate the MTTR one has to add the actual Mean Time To Repair with the Mean Time due to Logistics (MTL). The MTL is in many cases the determining factor.

When designing an appropriate rural telecom system for developing countries, with the required high level of availability, one has to take in account the special conditions mentioned before.

The equipment must be designed for high reliability. More reliable than ordinary telecom equipment, considering that the higher costs involved will pay themselves back through a reduction of maintenance costs and higher revenues. This high reliability can be achieved by careful selection of components, conservative design of circuits, use of inherently trouble free IC families, doubling of vital circuits, spacious lay-outing of printed circuit boards, the use of robust enclosures, low energy consumption design and last but not least careful thermal design, to avoid any need of air conditioning equipment.

Improving reliability alone is not enough. The MTTR has also to be reduced by all means.

MAINTENANCE FEATURES FOR IMPROVED MTTR

Remote monitoring and diagnostics. Nodes in a rural system, like the village switch, have to communicate with a maintenance center and will have to signal not only that something went wrong, but preferably also what module(s) is(are) causing trouble. A monitoring system that can detect detoriation of service well before a total breakdown is also of high value. This permits a maintenance crew to come well prepared at the site.

Controlled shutdown and auto-reconfiguration. A "well behaving" node should try to shut down step by step, possibly permitting operation with a limited grade of service through reconfiguration. In case of detected software malfunctioning the node will try to restart with internal or external reloading of software.

Auto diagnostics and repair assistance. During repair the built-in node software and supervision and maintenance module has to assist technicians as much as possible in clearing the faults.

A NEW APPROACH TO MAINTENANCE

A telephone system should be able to cope with several constraints imposed not only by its (future) users and the technical possibilities but also by the characteristics of the area of operation. Some of the particular constraints on telephone systems in rural areas that have to be dealt with are:

❖ continuity of service and maintenance in exigent circumstances.
❖ the unattended operation of the rural system nodes.
❖ the lack of skilled personel.

In order to offer a grade of service that does not differ too much from the urban areas special measures have to be taken to improve and simplify maintenance. Construction and maintenance operations should be supported by a modular design of switching nodes.
Moreover it should be possible to have access to reports on nodal functioning in a remote Operations and Maintenance Center (OMC). To enhance maintenance procedures these reports should not only instruct on what module has to be replaced but also give information on the future repair of the malfuctioning module.
Even more important are immediate service restoring actions taken in the occurence of a malfunction. As the majority of calls will have a local nature, a partially operational node is more preferable to the subscribers then a non-operational node.
Due to the unattended operation of the nodes and the lack of trained personnel all maintenance actions, except for the actual module exchange and repair, should be

executed automatically. This poses a requirement for monitoring, diagnosis and control facilities.

A distributed knowledge-based maintenance system can provide the necessary functions. Such a monitoring, diagnosis and control system [10] collects information on (mal)-functioning and diagnoses it in order to perform the before mentioned maintenance actions. This distributed approach to maintenance makes use of knowledge-based systems instead of conventional algorithms. This gives the advantage of easier changing and updating of expert knowledge in the case of nodal reconfiguration.

CONCLUSIONS

This paper proposes a telecommunication system for small villages in rural areas in developing countries (e.g. Africa). Further more it is shown that it is possible to provide a cost effective telecommunications solution for such a rural village using a fixed cellular radio system. This solution makes use of the spare capacity of an existing mobile cellular network of which the investment cost can be spread over a large amount of subscribers with a higher financial capacity than the average village dweller.

The grade of service that can be provided is not the highest available, but under the circumstances acceptable. In particular as some features have been implemented to guarantee fair access to the trunk lines. Through the design of the village station with a double radio transceiver, the availability of an outside trunk connection is improved substantially and also the access from a central maintenance center is better guaranteed.

The proposed system can readily be implemented on existing standard analog mobile cellular radio systems. However further study is necessary to look into possibilities of the use with digital mobile systems as promising cost reductions can be achieved.

ACKNOWLEDGEMENT

The author is grateful to Dr. Ramjee Prasad of Delft University of Technology for giving many interesting and useful suggestions and comments.

OoOoOoO

REFERENCES

[1]   Luis G. Romero-Font
      The use of Cellular Radiotelephone Networks to provide Basic Exchange
      Individual Line Telephone Service in Rural and Suburban areas ......, PTC
      1988, pp. 357-362.

[2]   P. Ashutosh and S. Kazeminejad
      Application of wireless access to telecommunication services in developing
      countries, ICUPC 1992, Dallas 1992, pp. 3.061-5.

[3]   M.D. Farrimond
      PCN and other Radio based Telecommunications Technologies for Rural
      regions of the world, 2nd Int. Conf. on Rural Telecommunications 1990,
      London, pp.99-104.

[4]   H. Hashemi
      Application of Cellular radio to telecommunication expansion in developing
      countries, 6th World Telecommunication Forum, Technical Symposium, 1991,
      Geneva, part 2,II, pp. 415-419.

[5]   I.A. Al-Kadi
      Integration of two telecommunication services: mobile radio and rural
      subscriber radio, EUROCON 1988, pp. 322-325.

[6]   M. K. Bin Harun and R. Omholt
      Malaysia Cellular System-Pioneer in Asia, Ericsson Review 1987,3, pp.151-
      159.

[7]   G.S. Raju and K.V.K.K. Prasad
      Introducing personal communications in the developing countries: Alternatives
      and Issues.

[8]   T.H. Chowdary
      Rural Communications, IEEE TENCON 1989, Bombay, 41.1.,pp 808-810

[9]   A. Diaz-Hernándes
      Rural Communications: A Comparative Analysis of Radio Technologies,
      Electrical Communication, 1st quarter 1993, pp. 91-96.

[10] L.G. Portielje and J.R. Westerveld
      A decentralised knowledge-based maintenance system for rural telephone
      exchanges, 2nd International Conference on Rural Telecommunications 1990,
      London, pp 141-145.

[11]   M.J. Morris and T. Le-Ngoc
Rural telecommunications and ISDN using point-to-mutipoint TDMA radio
systems. Telecommunications Journal, vol. 58, I/1991, pp. 33-39.

[12]   J.R. Westerveld and P.H.A. Venemans
A non-hierarchical rural network with decentralised control, 1st International
Conference on Rural Telecommunications 1988, London, pp 120-125.

[13]   J. Subramanian, K.G. Shanmugam and A. Mageswaran
Multi Access Rural Radio Telephone System (MARRTS), IEEE TENCON
1989, Bombay, 41.6., pp. 828-831.

[14]   G. Calhoun
Wireless Access and the Local Telephone Network, ISBN 0-89006-394-X,
Artech House, Boston-London, 1992.

PART V

COUNTRY STUDIES

Telecommunications and Development in Africa
B.A. Kiplagat and M.C.M. Werner, Eds.
IOS Press

CHAPTER XIX

THE HUMAN RESOURCES CHALLENGE - THE CASE OF CAMEROON

R. Maga
Director of CETCAM
(Centre d'Etudes des Télécommunications du Cameroun)

The Central African country Cameroon is known for its ethnic and geographical diversity, characteristics which have led this country to be labelled Africa in Miniature. Following a colonial status in the period between 1884 and 1916 under German rule, the country was divided between France and the United Kingdom and remained a protectorate until 1960. In 1961 both parts of the country were reunited and Cameroon became one independent state. As a result of this history, both French and English are in use as official languages. Cameroon measures 475,000 sq kilometres and has 12.2 million inhabitants. The rural population was almost 72 percent of the total in 1976 and at this moment around 62 percent. Population growth has been 2.9 percent annually.

THE TELECOMMUNICATIONS NETWORK

Today's telecommunications network is the result of patient work which never suffered major disruptions, despite the comings and goings of technical as well as political people. The development of the national network has evolved on the basis of a three phase plan, since the very beginning in 1960.

STAR NETWORK

The network inherited from before 1960 was a star shaped telephone network with decametric microwave connections, concentrated around two electromechanical switches which were located in the two principal cities in the country. Until 1968 expansion of the network was based upon this architecture. The installed switching equipment had a hybrid function, i.e. local switching and interlocal transit routing. For international connections, the subscriber of the Cameroon network was rather considered as an expatriate subscriber connected to a remote switch in France.

STRATEGIC PLANNING

From 1966 to 1979 strategic network development planning was put in place, which covered transmission, numbering, signaling and tariffing, as well as designing a hierarchical network structure in line with international practice. The two principal cities and the provincial headquarters could access basic telephone service. International traffic, automated for the majority of destinations, was secured through a transitting centre and a Standard A earth station.

Throughout the duration of the two development phases of this period, engineering and project management were carried out with almost permanent recourse to foreign expertise, especially as regards analysis and design. The Telecommunications Administration confined itself to setting general network development guidelines, as well as supervising implementation of agreed planning objectives which was done either on the basis of technical documents or inspection of completed projects. This supervision was usually undertaken by those technicians who would later be involved in the maintenance of those installations.

Notwithstanding all measures taken to ensure maintenance and proper functioning of the network, quality of service was at best mediocre as a result of reduced network availability. Human resources management and development were often regarded as of secondary importance. The failure to recognize the dimension of maintenance, as well as regulatory and legal challenges in the network development strategy, explain the gap between the good intentions of giving the country a good network and the results which are viewed with doubt by the users.

PLANNING GUIDELINES AND IMPROVED MAINTENANCE

The Government had requested a complete diagnosis of the telecommunications sector in 1979 - 1982, followed by a rehabilitation programme of installations and network segments. For this rehabilitaion programme, 25 international experts and 250 technicians were selected and hired by the Telecommunications Administration under ITU supervision. They worked throughout a period of 30 months, which had cost the Administration around US$10 million, including expenses for equipment.

In recognition of the importasnce of telecommunications in development, the Government decided to work out two key plans.

❏   The Planning Guideline for Telecommunications Development was created as a coherent development tool for networks and services, matching both needs and available resources;

❏   The National Maintenance Improvement Plan was adopted in order to formalize methods and procedures for maintenance of technical infrastructures, on the basis of fixed service objectives and of identified required human and equipment resources.

These plans were aimed at meeting the demand for telecommunications service at a level of 25 percent in 1988, 60 percent in 1995 and 90 percent in the year 2010. Self-financing should reach US$ 20 million, 70 million and 208 million along the same timescale, based on the CFA Franc parity against the French Franc after the devaluation of the former in January 1994. The above-mentioned startegies take into account the digitalization of the network, which is being undertaken on an 'island by island' basis.

TODAY'S NATIONAL NETWORK

Cameroon's national network has 64,000 operational main lines out of a total of 134,600 lines. 92,600 of these are digital lines installed between 1989 and 1991. The inter-urban trunk network consists of 5,500 analogue and digital microwave lines. Ten digital switches are in operation in the main cities Douala and Yaoundé, connected via fibreoptic lines. There are two earth stations for international traffic.

Basic telecommunications service started being introduced in the rural areas as early as 1986, on the basis of an experimental framework plan, while awaiting the results of a study to provide service to around 2,000 villages by the year 2005. Local staff is in charge of installing the rural systems, under the supervision of the supplier of the equipment. The supplier had previously provided training. Utilisation of indegeneous resources is being prsued more vigorously.

Cameroon has had an operational digital cellular radio network (GSM standard, 900 MHz) since October of 1993. An important characteristic of this cellular network is that its operation also serves rural telephony in the zone covered by the network.

MANAGEMENT OF THE NATIONAL NETWORK

Two independent entities are responsible for the management and operation of the national network: the international telecommunications company INTELCAM and the Direction des télécommunications, which is a directorate in the Ministry of Posts and Telecommunications. The latter is responsible for domestic telecommunications. Whereas INTELCAM staff is recruited on the basis of a private company charter, the 2,066 staff members of the Direction des télécommunications are hired on the basis of a charter applicable to all state employees. This means that telecommunications staff in the domestic operation is subject to the same rules and salary scales as staff in the Ministry of social affairs and advancement of women. This situation clearly has a negative effect on the sound management and operation of telecommunications services.

Following a restructuring framework inspired by the World Bank and the International Monetary Fund (IMF), it has been decided to have the two telecommunications entities managed by one single organisation. The legal status of this new organisation still has to be decided.

HUMAN RESOURCES

Human resources management is essentially recruiting, education and training including post-curriculum education, as well as personnel management.

INITIAL EDUCATION

Initial education for level 5 - 6 - 7 employees (ITU classification) must be obtained at higher education institutes outside of Cameroon. However, apart from the diversity in the educational programmes, there is the problem of establishing equivalent levels among certified senior personnel who has undergone training abroad. The creation and updating of an official reference list of educational institutes raises a delicate question.

The initial education of staff at levels 1 - 4 is provided locally at the Ecole National Supérieure des Postes et Télécommunications. The curriculum and recruitement in this school follows guidelines of the telecommunications Administration.

BREAKDOWN OF HUMAN RESOURCES

Two departments are in charge of staff. One is responsible for education and training of all posts and telecommunications personnel, on behalf of the government; the other is a staff bureau in charge of recruitment, career development and implementation of staff regulation, in line with rules applicable to all civil servants.

Table 1: Number of employees per 1,000 main lines

	1992	1993	1994	1995
Number of main lines	52,000	63,000	74,000	84
Number of employees	2,303	2,530	2,400	2,450
Employees/main line	44,3	37,3	32,4	29,2

The progressive improvement of productivity is a result of the fact that new recruitment in the public sector can only take place in exceptional cases, following a government directive which calls for a stop in recruitment and reduction of staff. On the other hand, the number of students enrolled in the Ecole Nationale Supérieur des P&T is decreasing year-by-year, since Cameroon is in the process of implementing a 'structural adjustment programme' as required by the IMF and the World Bank.

The reduction of staff per 1,000 main lines in Cameroon as shown in Table 1 is in step with trends all over Africa. Staff without technical qualifications represent 41 percent of the total workforce. Figures in Table 2 incorporate staff of INTELCAM as well, although they enjoy a different status than staff in the national telecommunications Administration. The negative growth of total wages is a result of retirement of a relatively large number of employees and a reduction of public sector salary levels imposed by the government. The average salary of 160,000 Franc CFA per employee per month in March 1994 (US$ 267) does not reflect the disparity between staff at the Directorate of Telecommunications and INTELCAM. The latter employs 13 percent of total staff but absorbs 34 percent of the wage total. This situation is complicating the necessary collaboration between the two entities. The high increase in operations costs is due to the devaluation of the Franc CFA in early 1994 and the fact that these costs involve many resources acquired from abroad, i.e. against 'hard currency'. The figure for turnover is on average only 60 percent of invoiced services rendered. The difference of 40 percent is caused by non payment by customers.

Table 2: Wages in relation to turnover

	1992	1993	1994	1995
Total wages (million Francs CFA)	4.490	4.481	4.093	4.093
Operations costs (million Frs CFA)	14.352	15.656	28.209	34.011
Turnover (million Francs CFA)	22.380	22.792	23.500	24.500
Wages % operations	31.28	28.62	14.51	12.03
Wages % turnover	20.06	19.66	17.42	16.71

note: the Franc CFA was devalued by 50% against the French Franc in January 1994

COSTS OF HUMAN RESOURCES DEVELOPMENT

The cost of initial education, available locally, is around US$ 6,700 per year per employee. The required sums for education abroad are on average US$ 92,000 per employee for a full cycle of five years. Technical education in relation to installation projects covers three main areas:

(a)   Theory related to equipment platforms (transmission, switching, outside plant, etc.). This first phase of training is provided at the initiative of equipment suppliers.

(b)   A selected group of technicians will go for in-depth training at training centres of suppliers and will later on take responsibility for workshops for installation of equipment.

(c)   Follow-up training in cooperation with suppliers is negotiated by contract.

Around 5 percent of the total value of contracts is generally devoted to training as explained above. The investment value of such training for the network can be estimated at US$ 11.250 million.

NEW CHALLENGES FOR HUMAN RESOURCES DEVELOPMENT

The telecommunications environment before 1980 was characterised by a relatively slow development of technology; well defined boundaries between different telecommunication services; a monopolistic market organisation; network growth in relation to economic development; and a weak interplay of national economies. Even in this relatively tranquil environment, the operation of the telecommunications network was almost an impossible challenge. At present, African countries must not only cope with profound changes in all the above-mentioned areas, but also with social and political changes inspired by democratisation processes.

We believe that in the area of telecommunications, the availability of quality human resources will determine the efficient utilisation of any other type of resources. This was illustrated by a recent ITU study, which showed that the loss of revenue caused by mediocre service quality came to 52 percent of total sales on average and 103 percent of annual investment. The maximum cases were 154 and 304 percent, respectively.

Equipment suppliers and the bigger network operators who determine quality and functionality of equipment, are restructuring their operations in order to be able to face the changes in today's telecommunications global market. Developing countries, particularly in Africa, have no choice but to adapt to changes imposed by larger players. The energies of telecommunications manufacturers are largely being absorbed by demands for digital networks, particularly Integrated Services Digital Networks (ISDN). As a result, they are forced to abandon production and support lines for analogue equipment. However, analogue equipment still makes up a very important part of operational installations throughout Africa. While our technicians have to continue maintain analogue systems, we have to develop new technical support capacity for the planning, installation and operation of digital networks. Industrialised countries have always been in a position to choose the optimum network development strategy, balancing analogue and digital technology. In Africa however, accellerated adaptation is imposed by global changes in the industry.

Furthermore, now that the commercial nature of telecommunications services is almost universally accepted, a new commercially oriented staff category has to be created. On top of the more traditional skills required by telecommunications operators, new expert capabilities are being solicited, such as marketing, law, information technology and others. While pressure on productivity is becoming stronger, with human resources the emphasis is rather on quality than on quantity.

CENTRE D'ETUDES DES TÉLÉCOMMUNICATIONS DU CAMEROUN (CETCAM)

The telecommunications Administration of Cameroon has continuously sought access to foreign expertise of engineering consultancy firms. The expansion and modernisation of the network could justify this activity and its integration in the operational framework of the Administration.

The Centre d'Etudes des Telecommunications du Cameroun (CETCAM) was created in January 1989. It's mission is to offer the Administration technical and economic programming elements, based on a widerange of technologies, in support of efficient operation of the network. CETCAM may also offer its services to third parties. The most important tasks of CETCAM are the following:

❖   Developing of Network Planning Guidelines

❖   Defining short and medium term investment plans

❖   preparation of calls for tenders for projects of the Directorate of
     Telecommunications

❖   technical evaluation of new equipment prior to installation

❖   maintaining technical documentation on relevant technology and equipment

❖   definition of structure and levels of tariffs on the basis of technical and
     economical data

❖   participation in defining sector policies

❖   participation in reorganisation or reorientation of educational programmes,
     including the Ecole Supérieure des P&T.

CETCAM consists of the following units:

❖   4 technical departments: local networks, transmission, switching, power supply
     and air conditioning;

❖   3 sections: engineering, economic analysis, documentation;

❖   telecommunications laboratory.

The telecommunications laboratory carries out specialised repair work on
malfunctioning modules. If necessary, the laboratory will evaluate and measure
transmission equipment. CETCAM has 43 staff, 19 of whom are engineers. The
Administration may call on external expertise in those cases where CETCAM lacks
capabilities.

CONCLUSIONS AND RECOMMENDATIONS

Management and operation of telecommunications networks is facing multiple
challenges, particularly in Africa. However, as stressed in "Telecommunication
Policies for Africa - The African Green Paper" (1), this should under no circumstances
bring us to a stop, but rather stimulate action.

Human resources are a critical factor in telecommunications development policies for
Africa, especially since almost all required equipment is imported. Equipment which is
in fact designed and manufactured without taking into account Africa's specific needs
and environmental conditions.

Evidence shows that well organized human resources have a multiplier effect on availability and therefore on the financial results of the operating companies. Professional education and training must be well adapted to the objectives of the public operators as the main beneficiaries of such programmes.

In almost all telecommunications Administrations in Africa, a number of negative developments have resulted in staff being under-utilized, poorly motivated and without obligations to achieve results. In many cases, this may lead to independent actions of 'intelligent' officers, often to the detriment of their employer. In African countries, stronger efforts in human resources restructuring are called for than ever before.

As recommended in The African Green Paper, we must invest in the development of human resources management, and ensure that procedures and regulation concerning personnel management are compatible with the commercial orientation of telecommunications services.

OoOoOoO

(1) Telecommunication Policies for Africa (The African Green Paper), second draft, November 1993. ITU Telecommunication Development Bureau (BDT), Geneva (Switzerland). The author of this article was one of the four regional African experts who contributed to this policy document.

Telecommunications and Development in Africa
B.A. Kiplagat and M.C.M. Werner, Eds.
IOS Press

CHAPTER XX

MOZAMBIQUE : NATIONAL OPERATORS SPARKS OFF PRIVATE PARTICIPATION (*)

Rui Fernandes
Chairman and Managing Director
Telecomunicaçoes de Moçambique (TDM)
Mozambique

The telecommunications sector in Mozambique is undergoing substantial changes following options taken by the Government to address basic policy issues regarding the provision of telecommunications services, the development and operations of the country's telecommunications infrastructure and the private sector participation.

The national operator Telecomunicaçoes de Moçambique (TDM) is an autonomous, fast growing and commercially oriented company. Focusing on its core business of providing telecommunications services, it has engendered a range of joint ventures involving foreign partners.

POLICY GOALS

In Mozambique, the primary telecommunications sector objectives fall under two main categories:

a.   the development of the telecommunications infrastructure and services, so as to make accessible to the population at large, both urban and rural;

b.   meeting the rapidly growing and changing diversified needs of the business community.

The strategic policies which the Government formulates to achieve these objectives are critical factors in determining the extent of success in reaching these objectives. The basic policies adopted need to take into consideration the specific characteristics of Mozambique's geography and demography, the levels of social organization and the economic and political orientation of the country.

THE TELECOMMUNICATION SECTOR

Institutional framework - In order to make the telecommunications sector play its part in the nation's socio-economic development under the best possible conditions, the Government of Mozambique has decided to reform the bodies constituting the telecommunications sector.

The decision forms part of the wider undertaking of reorganization and stimulation of the economy, a major component of which is the reform of the country's state-owned enterprises.

The goal of this undergoing process is, in particular, to:

a.   define the responsibilities of the Government for sector policy, strategy definition and regulation, and separate them from the operating responsibilities of the service providers;

b.   strengthen the public telecommuications operator, Telecommunicaciones de Moçambique (TDM)'s organization and management, corporate autonomy and accountability, and commercial capacity;

c.   increase the telecommunications coverage, improve maintenance and quality of service;

d.   enhance the sector's capacity to mobilize resources for the benefit of the Government (through taxes and dividends), and, as a result, improving its net contribution to the national state budget.

To achieve all these goals, the Government has approved the following laws during 1992:

❖   a telecommunications basic law;
❖   a decree establishing a regulatory body, the Insituto Naçional das comunicaçoes de Moçambique (INCM);
❖   a decree transforming TDM into a public company.

The interface and relationship between the government and TDM is now to be confined to these three levels:

❖   sector policy matters
❖   regulatory functions
❖   program contract

Telecommunications Sector's Entities - There are now three separate entities in the sector:

❖   the Government represented by the Ministry of Transport and Telecommunications;
❖   the Instituto Naçional da Comunicaçoes de Moçambique (INCM) which is the regulatory body;
❖   the service providers, with Telecomunicaçoes de Moçambique (TDM) being the main service provider.

SECTOR POLICY MATTERS

The object of the telecommunications basic law is to define the fundamental principles to be observed in the establishment, management and operation of telecommunications infrastructures and services.

The policy-making function is concerned with setting the objectives for the sector, how to achieve them and the principles and guidelines which will govern the day-to-day activities of the regulatory body.

The State has superintendence and inspection prerogatives in relation to the sector and the activities of telecommunications services providers, in accordance with the applicable laws and regulations. It lays down the strategic guidelines for the development of the national telecommunications system.

The State's prerogatives with regard to the superintendence, regulation and inspection of the sector also include the following:

❏   management of the radioelectrical frequency spectrum;

❏ representation in international inter-governmental organizations in the telecommunications field;

❏ definition of policies and overall planning of the sector;

❏ approving legislation and applicable regulations with regard to public use of the services;

❏ specifying standards and approving telecommunications materials and equipment, and defining the conditions for their connection to the telecommunications network for public use;

❏ granting concessions, licencing and authorizing the establishment and exploitation of telecommunications networks and services;

❏ inspecting compliance with legal and regulatory dispositions governing their activity by telecommunications services providers, and applying the respective sanctions.

❏ setting the prices and tariffs for telecommunications services, in accordance with applicable legislation.

REGULATORY FUNCTIONS

The regulatory function is concerned with the implementation of the sector policies.

The regulatory body is designated by the Instituto Naçional de Comunicaçoes de Moçambique (INCM), a public institute with legal identity, as well as administrative and financial autonomy.

The INCM is an institution subordinated to the Ministry of Transport and Telecommunications and its aim to assist the Government in coordinating, supervising and planning the sector concerned with communications for public use and in managing the radio-electric frequency spectrum.

The role of the Institute is to be limited specifically to:

❏ Formulation and interpretation of sector policy (provision of service, tariffs, liberalization, privatization, external resource allocation and mobilization);

❏ Licensing and regulating service providers and radio frequency spectrum users (issuing, monitoring, enforcing regulations and technical standards);

❏ Defining performance targets and indicators and supervision of the Program contract (access to services, quality of service, rate of growth and profitability);

◻   Handling relations with international and regional telecommunications organizations relating to treaties and policy matters.

PROGRAM CONTRACT

The activities of Telecomunicaçoes de Moçambique (TDM) shall be written into a Program Contract, signed for a period of three years between the Minister of Planning, the Minister of Transport and Communications, the Minister of Finance and the chairman of the Board of Directors of the company.

In the Program Contract, the Government and TDM agree on the strategic objectives to be achieved, on the policy with regard to tariffs, the size and type of investments to be made and on the financial, economic and social policy to be followed, including the parameters and targets of performance to be complied by TDM.

The Program Contract defines:

❖   the strategic guidelines for the company;
❖   the global objectives of tariffs developments for public services provided exclusively by the company;
❖   the main social, economic and financial guidelines for the company, in particular total salary commitments, investments and financing requirements;
❖   the principles underlying the application of profits;
❖   the criteria for assessing expected profits and nature of the respective indicators.

The current Program Contract covers the period from January 1, 1994 to December 31, 1996.

MARKET STRUCTURE

The issue of market structure is whether or not more than one service provider is allowed to provide the whole or part of the range of telecommunication services in open competition.

Competition in the provision of terminal equipment such as telephone sets, telex, facsimile machines and PABXx is generally beneficial, subject to properly designed and enforced technical standards, in order to promote diversity of supply and to relieve the main telecommunications service provider of tasks that others can do as well.

Similarly, competition in the provision of data transmission services and value-added services, merging telecommunications and informatics, is beneficial in promoting diversity of supply of services.

Where competition is permitted, the established service provider faces competitive pressure to offer improved service and prices to its customers.

Under the new telecommunications basic law:

❐   TDM will operate, on a monopoly basis public telecommunications network, consisting of the fixed systems of connections to subscribers, the transmission network and switching, concentrating and processing equipment, to provide the basic services of telephone, telex, fax and also cellular mobile radio to subscribers.

❐   Data transmission, value-added services and complementary services will be open to competition subject to decisions and regulations to be issued by the regulatory body.

The infrastructure that covers the basic telecommunications services is part of the public domain.

TARIFFS

Tariff policy is of primary importance for a telecommunication organization such as TDM, which operates along the lines of a business enterprise. Tariffs structure and levels largely determine the revenue generating capacity of TDM from its operations.

The basic tariff policy is derived from the sector's financial policy which calls for TDM's operation to be profitable. Current basic policy is hence that tariffs should be set so that:

❖   gross revenue generated should cover costs;
❖   rate of return on investment should enable self-sustained growth;
❖   revenues should be stable despite the uneven profitability of the different types and categories of services;
❖   revenues should also be unaffected by changes in the national economy environment, in order to ensure the planned, self-sustained and steady growth of the network and services.

The second consideration is that the basic tariff policy, although targeted to achieve financial objectives, needs to be reconciled with social, economic and political objectives and Government policies in other sectors. These relate to issues such as the need for uniformity of tariffs nationwide and the extension of services to rural communities.

Thirdly, the basic telecommunications services are operated under a monopoly nationwide. In a market where supply rarely satisfies demand, there is no market mechanism to press down and to regulate prices. Therefore, the Government has the inevitable regulatory duty of ensuring that tariffs reflect costs.

Under the new telecommunications basic law, the Government will fix the tariffs for the services operated as a monopoly. Tariffs for all other services where competition applies will be fixed by the service providers themselves.

As a matter of general interest, Government has agreed a tariff policy which allows for tariffs to be adjusted regularly in order to protect the real value of tariffs in a high inflation environment.

CREATION OF TELECOMUNICAÇOES DE MOÇAMBIQUE (TDM)

Institutional Framework - Traditionally, the postal and telecommunications services have been part of the Government structures, without a clear distinction between policy, regulatory and operational functions.

In line with the development mentioned earlier on, Mozambique has separated the postal and telecommunications services from the government structure in 1981, creating two distinct state enterprises.

Law 17/91 of August 3, 1991, introduced a new legal system for state enterprises, thus originating profound alterations in the management of companies with state capital, and requiring the modification of the articles of association, which are to be transformed into public companies.

The Empressa Naçional de Telecomunicaçoes de Moçambique, E.E. (state enterprise) created on January 1, 1981, was hereby transformed into a public company with the designation of Empresa Naçional de Telecomunicaçoes de Moçambique, E.P. (public company) from January 1, 1993.

Telecomunicaçoes de Moçambique, E.P. (TDM) succeeded automatically and totally Telecomunicaçoes de Moçambique, E.E., maintaining its corporate identity, retaining all its property and assuming all the rights and obligations derived from the acts or contracts celebrated at the time of the transformation.

Telecomunicaçoes de Moçambique E.P. is a public company, a corporate entity with administrative and financial autonomy and its main objective is the planning, installation and operation of the national and international public telecommunications service. The public telecommunications services includes the public telephone and telex services and is provided on an exclusive basis.

Telecomunicaçoes de Moçambique E.P. can also undertake commercial, industrial and financial activities related to its main activity when duly authorized by the Minister of Transport and Communications and Minister of Finance.

ORGANIZATION

An organization development process was set up in 1989 with the aim of creating a business oriented services company, managed according to modern management principles and starting a process of decentralization and delegation of authority for clearly defined responsibilities and its financial results.

Successive changes of TDM's internal organization have been undertaken during the least years and a major change has been implemented in 1992, on the verge of the transformation of TDM into a public company.

During this period, the evaluation of the new organization has been done aiming at the necessary modifications, as well as the preparation of a manual for a regular system of organization development.

The internal organization chart now features four Directorates below the Chairman and Managing Director:

❖ Operations;
❖ Finance and Administration;
❖ Planning and Engineering;
❖ Telecommunications Districts.

The heads of theses directorates are also Board Members and seven functional directors report to the Operations, Finance and Administration and Planning and Engineering Directorates.

The Telecommunications Districts Directorate has nine operations areas, corresponding more or less to the administrative divisions of the country. Maputo City and Maputo province come under the responsibility of both Operations and Finance/Administration Directorates.

MANAGEMENT

A Board of Directors (BOD) assisted by an audit Council constitutes the top level of TDM's organization.

The BOD establishes the TDM's internal policy and supervises and coordinates TDM's activities in line with corporate objectives. It establishes staff rules and regulations, appoints and terminates executive directors, approves multi-annual and annual corporate plans, issues annual reports and statements of accounts, proposes the utilization of profits, approves disposal and acquisition of assets, and submits proposals for the formulation of sector policy or legislation issues to the Ministry of Transport and Communications, where and when required.

The BOD has seven members, including the Chairman, of which four member are proposed by the Chairman and appointed by the Minister of Transport and Communications, one by the Minister of Finance and one by TDM's empoyees. The Chairman is appointed by the Council of Ministers and is also the Managing Director. The Chairman and the four Board members appointed by the Minister of Transport and Communications have executive management responsibilities for one or more functional areas, with a tenure of three years.

The Audit Council is composed of three members, all appointed by the Minister of Finance for a five years period. The Council members are remunerated by TDM on a fee basis. The role of the Audit Council is to monitor the financial operations of the company and to audit it in compliance to laws and approved implementation plans. It also determines criteria of asset re-evaluation and depreciation rates, examines and issues opinion on annual reports and statement of accounts, evaluates financial performance and operational efficiency, draws attention of the Board to any preceived shortcomings and advises the Board on any relevant issue, when requested.

The management of the company is conducted in accordance with State economic and social policy, and principle of economic calculation which can be fixed and controlled objectively for its various functions and activities. Therefore, the management of the company is observing the following principles:

❐ short and medium term economic and financial objectives clearly fixed in the Program Contract celebrated with the Government;

❐ principles of economic self-sufficiency, except when the State, for political reasons, imposes tariffs below normal or fixes social objectives which are not economically profitable for the company.

❐ the price policy approved by the Government for the services in which the company has exclusivity;

❐ a salary policy which takes into account salary scales in the national labor market, signing collective labor agreements aimed at creating social harmony and evolution of salaries based on productivity;

❐ the assurance of economic and financial rates of return on both existing investments and new ones.

❐ new investments are subject to entrepreneurial decision criteria, as regards to rates of return, capital recovery period and risks, except when other criteria have been agreed with the State;

❐ secure adequate financial resources for the assets to be financed;

❐ a financial stucture compatible with the profitability and the risk level of the activity;

❏ management by objectives, based on decentralization and delegation of responsibility;

❏ ensuring constant productivity increases while minimizing production costs.

Whenever the company is obliged to apply tariffs lower than normal or is obliged to pursue social objectives which are not economically viable for the company, the State shall provide a budget subsidy to compensate for the costs not covered through its own income.

STRATEGIC OBJECTIVES

The Government and TDM agree, in the Program Contract, on the strategic objectives to be achieved, tariffs policy, scale and type of investments, and on the financial, economic and social policy to be followed including on the parameters and targets of performance to be complied with by the company.

The strategic objectives in the Program Contract for the three year period 1994 - 1996 and the focus on:

❖ development of the network to satisfy demand for services;
❖ raising the quality of services being offered to the public;
❖ maintaining the financial profitability of operations;
❖ consolidating institutional development aimed at creating a high productivity and motivated work force.

A three-year Corporate Plan 1994-96 has been prepared, defining the strategy and elaborating the activities to be carried out in order to achieve these objectives. A yearly plan of activities and implementations is based on this plan.

In line with these strategic goals, specific objectives have been fixed in the functional areas of network development, operations, finance and supply, and human resources. Performance target values are set for these areas in the Program Contract with the Government.

NETWORK DEVELOPMENT

Network development is the task of increasing service penetration and access to service using appropriate technology. The qualitative objective is to replace all analogue switching equipment by digital switches and all new equipment will use digital technology. Quantitatively, specific targets have been set for the expansion of the network.

240

OPERATIONS

The goal of the Operations Plan is to maintain an acceptable standard of quality of service both from the technical and commercial point of view to keep the customers satisfied. It also aims at the optimum utilization of the telecommunications infrastructure. The Operations Plan has been elaborated indicating the volume and the types of service to be provided, the required resources taking into account ongoing investment. Relevant performance indicators and targets for the three-year planning period area included.

Main indicators of quality of service, both for telephone and telex services, are fault rate, fault repair rate, delay time in fault repair, installation delay time, billing error rate and delay in resolving billing errors. Network utilization is monitored on the basis of percentage of equipment utilization and the rate of completed calls for urban, trunk and international traffic flowing through the network.

Finance - The financial strategic objectives are the following:

❖ to provide sufficient funds for the investment program through internal generation and mobilization of external financing on attractive terms;
❖ to ensure the profitability of business operations;
❖ to ensure a high rate of return on investment.

Strategic actions in the financial domain currently being undertaken include:

❖ to strengthen the financial and supply management functions through reorganization and upgrading of skills;
❖ to tighten financial control to reduce costs and a more aggressive commercial position to increase revenues from sale of services;
❖ to increase the rate of collection of revenue generated from sale of services;
❖ to seek better financial terms for external finance;
❖ to institute a tariff adjustement mechanism that is flexible and cost sensitive;
❖ to exploit opportunities for investment in activities related to TDM's line of business;
❖ to elaborate a series of financial plans, the budget (one year and very detailed), the Three-Year Plan (less detailed but covering all costs and revenues) and the Corporate Business Plan (long-term strategies with contingency plans for a range of possible futures).

Main financial performance indicators, to be included in the Program Contract, are rate return on assets, operating margin, debt service coverage, debt equity ratio, self-financing ratio and number of days in debtors arrears.

HUMAN RESOURCES

Human resources management and development aims at:

❖ increasing staff productivity;
❖ creating better conditions for job satisfaction.

A number of measures are being taken currently in the area of human resources management to achieve these goals. These include:

❖ improvement of the system of management of human resources;
❖ re-organization of the human resources management unit;
❖ development of a human resouces plan;
❖ institution of a system of career structure and development;
❖ development and implementation of administrative procedures, staff rules and regulations;
❖ recruitment of new staff with higher education and reduction of staff in the lower categories, while controlling overall staff growth rate.

The number of staff per 1,000 direct exchange lines (DEL's) is the performance indicator used to measure staff productivity.

The strategic goal of the training function is to ensure that TDM staff has the specific skills and work attitude needed. Initially, priority was given to training of technical staff at the basic skills level. This goal has already been achieved and local training capability has been built up to a satisfactory level.

Current objectives for training are:

❖ widen the scope of training to cover adequately administrative areas;
❖ develop a programme for higher level management staff;
❖ upgrade the technical training programme to produce medium and higher level technical staff.

INSTITUTIONAL DEVELOPMENT PROCESS

There have been marked achievements in the overall institutional development of TDM in recent years. Organizational development and changes in the systems of management and working procedures are being introduced in all areas and levels. Some of the main developments in the principal functional areas are presented below.

PLANNING

The planning function has two categories:

❖ Corporate planning;
❖ Network development planning.

CORPORATE PLANNING

A corporate planning unit was established recently as a support function of TDM executive management in:

❑ setting clearly defined short and medium-term strategic objectives for the company;

❑ coordinating plans and ensuring that initiated functional level activities are properly consolidated, e.g., as regards network development, operation and maintenance improvements as well as the development of financial and human resources;

❑ setting performance objectives and performance indicators at corporate and functional levels;

❑ monitoring and evaluating performance using a computerised Management Information System;

❑ assisting top management in the decision making process by providing timely information on performance analysis.

NETWORK DEVELOPMENT PLANNING

For several years, TDM did not have a distinct planning unit and was dependent on external assistance for network development planning and engineering. This situation is now corrected.

❖ the Planning and Engineering function is established;
❖ a Master Plan covering the period 1991 to year 2010 has been prepared in collaboration with the ITU;
❖ in-house planning and engineering capabilities are upgraded;
❖ technical assistance to further strengthen the Planning and Engineering Directorate is being arranged.

OPERATION AND MAINTENANCE

TDM has devoted considerable resources both of its own and from external assistance in plant operation and maintenance which has contributed to quality of service.

Priority was given to the establishment of maintenance support facilities in various specialized equipment areas. A Central Laboratory is in operation since 1989, equipped with modern testing equipment for electronic cards for digital switching equipment. A computerized local network management system is in operation in Maputo and surrounding areas. The local network maintenance facilities are upgraded during the last couple of years. In the area of network management, a system for centralized control and management of the switching and transmission system is in operation. TDM is now considering acquiring a TMN system for centralized control of the entire national network.

COMMERCIAL OPERATIONS AND MARKETING

TDM has already initiated major changes in this area:

❖ the billing and collection processes are computerized in the Maputo area;
❖ the credit control system is reorganized and tightened;
❖ new and more suitable premises are being prepared to welcome the clients;
❖ an integrated computerized system for service request and subscriber record management, local network and fault management is introduced;
❖ organization and management of the commercial activities is being streamlined at all levels, central, regional and local;
❖ procedural manuals are being prepared and staff given training in the new market-oriented systems and procedures.

The aim is to change the overall orientation of the commercial operation of TDM and instill a more aggressive marketing spirit. This entails introducing changes in staff attitudes and skills as well as in the organization and management of commercial operations.

FINANCIAL AND SUPPLY MANAGEMENT

There has been significant improvement in the area of financial management since 1989:

❖ a commercialized accounting system is introduced;
❖ the yearly balance sheet and profit and loss statement is being produced regularly;
❖ a computerized accounting sytem for the consolidated accounts at the head office level is installed.

The financial and supply management still require changes, in view of the transformation of the company from a parastatal to a public corporation with financial autonomy required to operate like a business enterprise.

TDM's accounts are being brought in line with commercial practice. An opening balance sheet is prepared as off January 1, 1993 and internationally recognized external auditors are appointed. Discussions are under way with Government on restructuring the financial position of TDM and resolving various components of its financial relationship with Government for services rendered.

A thorough review is carried out of the deficiencies relating to organization and management, systems and procedures, as well as skilled manpower needs in the financial and supply management areas.

A consulting firm is now engaged to help reorganize the Finance Department activities and implement systems and procedures in the areas of financial management, general accounts, cost accounting, budgets, revenue accounting, tariffs and international accounting. A similar exercise is being carried out in the Supply Department. This task is projected to be completed by 1995.

HUMAN RESOURCES MANAGEMENT

TDM has a total work force of 2,580 employees as of September 30, 1993. The average educational level of the staff has been quite low and a major impediment to improving productivity. TDM is continuing to recruit new staff with higher level of education, in particular, university graduates, and to reduce the lower level staff in order to improve the composition of the staff. TDM also supports part-time attendance of its staff at the university and other high-level educational institutions. Simultaneously, TDM has taken several measures to improve the conditions of work for its employees. These include:

- ❖ a comprehensive set of rules and regulations governing personnel administration;
- ❖ job analysis and elaborated career structure for different categories and skills;
- ❖ a better salary scale and improved employment conditions;
- ❖ a vigorous recruitment program has transformed the composition of the manpower: the number of middle level staff increased by more than 130%, the number of university graduates increased by almost 700%, and the size of the existing staff has been reduced by about 8% in a relatively short period of time.

HUMAN RESOURCES DEVELOPMENT (TRAINING)

Training policy has been consistent with overall personnel policy and the human resources management system. In line with its strategic human resources objectives, parallel measures have been taken to upgrade the quality of TDM staff in all areas.

❑ the intake level of trainees has been raised from the present relatively low junior high school level to grade 11 or better with expected benefits of training time and cost reduction and higher quality output;

❑ a modern physical infrastructure is built and put into operation;

❑ a training management system (methods, procedures, administration of course design, production, reproduction, delivery, evaluation) is created;

❑ Mozambican staff are fully in charge of the operation of the training complex;

❑ the Training Institute offers a full set of courses on technical subjects at the basic level and some at medium level.

The Training Institute is now widening the scope of the courses offered and is including courses in administrative areas. In addition, the training program is giving special attention to developing management skills of TDM's higher level staff.

INVESTMENT PROGRAM

The telecommunication network of Mozambique remained unchanged for the first five years since independence in 1975. Soon after its establishment in 1981, TDM set out to prepare a long-term development plan and a medium-term investment program. A master plan was prepared with the ITU in 1982. The first investment program entitled the Telecommunications Development Project (TDP) was formulated immediately with the assistance of consultants, covering the ten-year period from 1982 to 1992.

Resource mobilization for the TDP took considerable effort and time with the implementation starting in 1986. By the end of 1989 Phase I of the TDP was completed. The implementation of the TDP resulted in a major transformation of the telecommunications network in Mozambique. The size of the network was doubled. New digital technology was extensively introduced. An extensive domestic satellite communications network was established. The international and trunk services were automated. Local networks were rehabilitated and extended in major towns. During the period 1989-1993, the expansion of the domestic satellite communications network continued and the introduction of the rural communications started in all provinces of the country.

FIVE-YEAR INVESTMENT PROGRAMME - 1994-1998

TDM is now preparing a five-year investment programme for the period 1994-1998 based on the 1991 Master Plan prepared with the ITU. This programme will cover the digitalization of the national and international satellite network, the implementation of Phase II of TDP, the expansion of international facilities including the international exchange and a new earth station, the expansion of the rural network, the improvement of the network in the DADC region, and the introduction of a maritime system and a cellular mobile network in the southern part of the country. In 1995 the switched network will have reached a capacity of 100,000 lines and network digitalization will have reached 75 percent.

Rigorous economic and financial appraisal of proposed new investments will be carried out and a financing strategy will be prepared to allow TDM to secure sufficient financing on acceptable terms at the required time. This strategy will include negotiations with the Government and identification of providers of finance, as well as interfacing with the procurement strategy.

As part of the programme for mobilizing funds to meet planned investments, a seminar on Mozambique Telecommunications Development and Management was held in Maputo in early April 1994.

The objectives of the seminar were three-fold:

❖ to keep the current financiers informed about the implementation status of the Operational Management Plan (OMP)'s projects;
❖ to inform all participants about telecommunications development in Mozambique since the last seminar (May 1992);
❖ to present TDM's development programme and secure commitments to finance it.

Participants were international financial institutions, development agencies, telecommunications services providers, telecommunications consultants, telecommunications suppliers, the Government of Mozambique, the regulatory body (INCM) and TDM's main business customers.

INVOLVEMENT OF THE PRIVATE SECTOR

The telecommunications basic law has created options for the involvement of the private sector in telecommunications. According to the law, direct or indirect participation by foreign individuals or collective bodies in the share capital of complementary or value-added telecommunications service providers can be up to 50 percent. However, the Government can fix other percentages by decree, on a case by case basis, in accordance with the evolution of the market in telecommunications services.

On the other hand, in spite of TDM being wholly-owned by the Government of Mozambique, existing legislation empowers TDM "to carry out commercial, industrial and financial activities related directly or indirectly to its main activities".

TDM is cautiously exploring the benefits of private participation in the telecommunications sector. Private participation is being introduced in the form of joint ventures in a number of value-added and peripheral areas such as: editing and printing of telephone directories, local network construction, supply of terminal equipment, maintenance of vehicles, electronic security engineering, data services, telecommunications consultancy and civil construction.

exhibit 1

> ✧ TDM Joint Venture Companies
>
> ✧ Listas Telefonicas: editing of telephone directories
>
> ✧ Teledata: data communication services
>
> ✧ Televisa: local network construction, country-wide
>
> ✧ Teleserve: marketing of terminal equipment and paging systems
>
> ✧ Teleconsultores: telecommunications consultancy with SvedTel (Sweden)
>
> ✧ Telealarme: electronic security engineering
>
> ✧ Sogitel: civil construction
>
> ✧ Servisa: maintenance and leasing of vehicles

TDM is a fast growing organisation with telecommunications operations as its core business. The joint ventures with foreign partners ease the financial and management burden on TDM for these services and enables it to devote available limited resources to its principal business operations. The indirect benefits of this policy, apart from the financial dividends to be expected eventually, is the explicit costing of support services and the increased cost consciousness in internal and external resources utilization. TDM and external partners each have 50 percent equity in the joint ventures. By contract, management of the joint ventures is fully independent of TDM. These management contracts are for three to five years.

TDM's CHALLENGE AND OBJECTIVES

TDM's challenge, therefore, is to provide a wide variety of high quality services at a profit. This challenge is reflected in TDM's recently formulated objectives. We believe that these objectives will be achieved more quickly and also that existing and future investments will be better utilized with the establishment of joint-venture companies with private partners.

In effect, through these joint-venture companies, we are introducing a measure of privatization to the sector and, as a result, TDM will be better placed to deal with any competition which already exists or may emerge in the future.

INTERNATIONAL PRIVATE PARTNERS

TDM decided, at an early stage, that suitable international partners were essential for the success of the joint-venture companies. Over the years, TDM has dealt with many foreign consultants, service providers and other companies. Some of the problems encountered are related to the application of experiences from one environment not being directly and readily applicable to its new proposed environment. Therefore, in its selection of international partners for the joint-venture companies, TDM has become careful to ensure that candidates have previously collaborated with and worked successfully in Mozambique.

Objectives of the Joint-Venture Conpanies - The objectives of all TDM's joint-venture companies include offering high quality services and being recognized first in Mozambique and later in the Southern Africa region as the best alternative for providing services in their respective areas. With the establishment of these companies, a pool of directly applicable region-based experience will be built up and TDM intends to actively market these skills and experience within the region.

LAW FOR FOREIGN INVESTMENTS

The Government has been promoting the establishment of joint-ventures with foreign partners and a new law regulating direct foreign investments was recently approved. The objective of the establishment of such companies is to better serve the market by introducing the required know-how and capital.

Mozambiques incentives to the establishment of joint-ventures include:

❖ protection of property rights;
❖ tax exemption arrangements;
❖ exemption from payment of custom duties;
❖ repatriation of profits by the foreign investor.

CONCLUSION

The development of the telecommunications sector in Mozambique has undergone a process of evolution in recent years. The Resolution ATDC-90/RESI of the African Telecommunication Development Conference (Harare, 1990) concerning the African Information and Telecommunication Policy and Strategy recommended that African Governments:

☐   separate the regulatory and operational functions of telecommunications;

☐   separate the postal and the telecommunications management functions;

❑ merge domestic and international operating entities;

❑ give the telecommunications operating entities a large financial and managerial autonomy, along with the appropriate accountability;

❑ create a consultative mechanism that allows the involvement of users and other parties concerned in the improvement of the efficiency of the sector;

❑ establish in each country a national high-level multi-ministerial coordination committee to study and propose to Government appropriate national information and telecommunications policies that will cover sector regulation and operation;

❑ include in the national coordination committee representatives of the Office of the Head of the Government and of the Ministers responsible for telecommunications, information, finance, planning, agriculture/rural development, trade/industry and a representative of the users/clients.

Despite the war that has ravaged Mozambique until 1992, the country has already implemented the first five of these seven recommandations (two of them will before 1990) and is in the process of implementing the remaining two. Government policy in telecommunications has resulted in steady increase of basic telecommunications access facilities and services.

Meanwhile, the telecommunications basic law and the law for foreign investment have provided an appropriate business climate for the private sector and the mobilization of resources for investment and development of the infrastructure.

Through its institutional development process, Telecomunicaçoes de Moçambique (TDM) has been able to modify its business structure to become a market-oriented enterprise. TDM is in a strong position to act as a catalyst for economic growth and to serve as an institutional model for economic recovery.

OoOoOoO

(*) Text based on presentation World Bank Conference Telecommunications and Economics Prosperity, Abidjan (Côte d'Ivoire), 21-25 February 1994.

Telecommunications and Development in Africa
B.A. Kiplagat and M.C.M. Werner, Eds.
IOS Press

CHAPTER XXI

GAMBIA: IMPACT OF RURAL TELECOMMUNICATIONS ON AN OPERATOR'S BALANCE SHEET

Bakary K. Njie
Managing Director, Gamtel Company Limited
Gambia

ABSTRACT

This paper describes the strategic importance of the judicious phased programme of telecommunications development and modernisation in the rural areas with the corporate objectives and goals of improving the overall low national telephone density and also enhance revenue generation through aggressive marketing and business plans. Cautious investment outlay directed towards Regions and/or Divisions of commercial, agricultural and administrative importances are featured prominently in this paper as they reflect the operator's well conceived strategies and expectations of optimistic profitable delivery of services to the rural population.

INTRODUCTION

Rural telecommunications has a long history of countless debates and arguments as to its modality and definition. It can be defined, in my view, as providing telephone network and services to areas where there are few people, unduly separated and difficult to reach, compared to urban cities, towns and their suburbs. Usually, from the operator's point of view, we talk about rural telecommunications providing services to remote towns and villages which are uneconomic due to militating factors like high installation and maintenance costs, low returns and subsidised tariffs which do not reflect the current cost and market situation.

I intend, however, to demonstrate that rural telecommunications, if viewed in a cautious and businesslike manner and tailor made to the requirements of the user group, can be turned into a profitable venture. This can only come by if the will and determination to succeed are present, with prudent proper financial and managerial discipline.

CHOICE OF TECHNOLOGY

Rural telecommunications, as mentioned above, are composed of a physical network of exchanges, a transmission system and external plant. Most rural networks in Africa are characterised by frequent breakdowns, inefficient service, poor management and obsolescence of the technical equipment. To improve service quality in the rural areas, many African administrations are now in the process of rehabilitating and/or completely replacing their rural equipment with the latest available technology, as the national telecommunications operating company GAMTEL Limited has just just done in the Gambia with the installation and commissioning of a backbone fibre optic link along the length of the country.

However, the choice of the level of technology appropriate to various portions of the rural network is a very important decision for administrations. There is always the temptation to specify nothing but the most advanced technology available. Sometimes this may be the best solution, but sometimes not because of violent environmental conditions, pessimistic effective calling features of equipment and staff inadaptability in managing the new technology.

Some useful thoughts on the topic of appropriate technology were set out by the Deputy Secretary General of the ITU Mr. Jipguep in the ITU Telecommunications Journal, Vol.551/1988. After analysis of some of the perils of attempting to adopt "appropriate" technology, an extract from this article reads:

"I feel that the following are among the prime considerations for appropriate technology in telecommunications for developing countries:

❖ the technology must be technically appropriate to meet the present expected future demands;

❖   the implementation of the technology must be economically viable;
❖   the technology must have high reliablity and maintainability;
❖   it must call for a minimum of power and space requirements;
❖   it must be able to withstand the prevailing climatic environment."

The last point, regarding the climatic environment, is worthy of special mention as a mitigating factor which can inhibit efficient performance of the rural system.

The choice of technology will depend on a number of factors, as depicted above, in addition to traffic capacity and finally the cost and availability of financial resources. These precautious will ensure optimal utilisation, availability and quality of services in the rural areas and also consolidating the user's confidence in the system's reliability, which will regenerate demand to be readily met by the ultimate capacity of the system, as conceived in the earlier stages of planning and traffic forecast.

CHOICE OF SERVICE GROWTH CENTRES

One of the distinctive features of any rural settlement in Africa is the diverse characteristics of some centrally located towns and villages, which act as converging points for commercial and administrative purposes on one hand, and also convenient access points to essential services (medicare, education, telephone, local government) on the other, in any regional divisional area.

The need to maximise revenue by covering those areas with the highest forecast traffic early in the evolution of the network cannot be over-emphasised. This can be marginally achieved by careful design and choice of the backbone network structure around these centres, whilst catering for future-proof capacity against possible developments and suppressed demand in the localities.

There must be prudent restrictions on the scope of the project or investment because of the wider implications on the financial performance of the administration and indeed the country at large. Hence the need to confirm the traffic and revenue assumptions which must be translated into judiciously phased programmes of service extensions, based on demand and requisite commercial considerations for enhanced revenue generation due to rapid traffic growth.

Many governments in Africa see reliable telecommunications network for rural areas as vital to the economic development of the whole country. No matter how distant a given service centre may be from surrounding settlements, the rural population will always access it by available transportable means, once they are assured by circuit availability and quality of service.

Sir Donald Maitland, the chairman of the Independent Commission on Worldwide Telecommunications Development, in addressing a Conference on rural telecommunications in May 1988 in London, England, said:

"The advantages an extension of the network will bring to inhabitants of rural areas, whether they be in Australia, Ethiopia or India, will be broadly similar. The rural economy will be stimulated, employment opportunities will increase, migration to urban centres will decrease, transport costs will be reduced, marketing will be more efficient, emergency services will be more readily available, the sense of isolation will be diminished and the quality of life and national cohesion enhanced."

It is obvious that the rural services should initially be geared towards population growth centres of commercial and administrative importance, which will gradually spread the "telephone culture" to the surrounding rural population for greater social interaction and consolidation of family values and virtues so unique in the African context.

EFFECTS OF OUTLET CHOICE ON THE BALANCE SHEET

If any investment is undertaken in the rural areas without due consideration for the consequences, especially on the operator's cash flow, then both the investment and even the operator's own survival will be at risk. As mentioned earlier, there must be prudent restrictions on the scope of the rural project.

Various choices relating to suitability and size of the equipment and the targeted population and their commercial centres are important business decisions that influence increased volume of service connections and traffic which will have a positive impact on the balance sheet.

What is a balance sheet? A balance sheet is just an indication of the stock of assets and liabilities of the operator at a point in time. It is also a statement of financial position. As a statement of financial position, any retained profits and investment made during the course of the year will have a significant impact on the operator's balance sheet at year end. This means that the profits will increase the balance sheet and investments, if funded by borrowing or equity contributions, will also increase the balance sheet. Therefore, rural telecommunications operations, which require vast investments, will have considerable impact on an operator's balance sheet.

Having defined rural telecommunications and balance sheet, how do we go about identifying and measuring the impact? The relationship between rural telecommunications investment, net profit and their impact on the balance sheet has already been demonstrated.

My discussion on the impact of rural telecommunications on an operator's balance sheet will be based on GAMTEL's experience over the past ten years of its establishment (1984) as a limited liability company responsible for all telecommunications matters in the Gambia, both national and international services.

GAMTEL'S EXPERIENCE

It is obvious that the cost of the telephone service is higher in rural areas, where for instance distances are larger and where the economic advantages of large-scale production do not exist, than in the urban areas. It seems acceptable and reasonable that the urban subscribers should subsidize the rural customers in order to reduce the charges for the latter.

Another reason for this subsidising is that it facilitates the expansion of the network to remote parts of the country, which also gives added value to the telephones in the urban areas.

Telecommunications investment is capital intensive and GAMTEL's mission is not a telephone for everybody, but close telephone access by everybody in the Gambia and good quality of service for greater customer satisfaction.

The rural telecommunication project (phase III) only came about in the Gambia after successful development of the urban network which assured sustained yearly increased profitability of GAMTEL and in the building up of healthy reserves for future development undertakings.

This strategy made a greater impact on the rapid growth rate of the company in terms of the balance sheet, profit and loss account and human resources development. The figures in appendix 1 are taken from the balance sheets in the Final Accounts of the company.

For techno-economic viability it is more appropriate to consider a group of villages or some unit area as a basic unit and design a network for connecting rural villages in the group to a group centre. By such an integrated approach not only all the villages at the group get service access to the group centre, but also the entire network can be considered as a single network in which optimum utilisation can be made of transmission systems and switching nodes. The selected group centre could be an administrative headquarters or an important commercial town.

By considering all the villages simultaneously, it will be possible to economize the network planning by choosing in the initial stages the appropriate type of transmission links and also by fixing the optimum number of switching nodes.

Village telephones which have low calling rate and are spread over an area need transmission links for access to the switching node at the group centre. Providing individual land dedicated transmission links to every village will obviously be costly and cause undue burden on the financial stability of an administration.

With the above concept in mind, GAMTEL's phase III rural project was conceived and implemented with detailed revenue/expenditure projections as shown in Appendix 2. The cross subsidy from urban networks is also shown as incoming traffic from urban areas. The underlying management commitment at the birth of phase III was never to erode the financial stability of GAMTEL and also to maintain overall 15% return on capital employed prior to depreciation, interest and tax. The projections showed that this will be achieved even with the rural project investment. It is important to note that the scenario includes incoming revenue, which is a most realistic approach in evaluating rural investment projects.

The latest balance sheet (1993/94) is an encouraging indication that the rural phase III project will enhance GAMTEL's profitability and narrow the gap between urban and rural telephone density.

OoOoOoO

APPENDIX 1: BALANCE SHEET GAMTEL

Balance Sheet GAMTEL (1,000 Dalasis)		
Description	**March 93 (actual)**	**March 94 (budgeted)**
Fixed assets	318772	362919
Investments	1300	11300
subtotal	1300	374219
Current assets		
Stock	28931	22000
Subscribers debtors	20421	30860
Other debtors & debit	36844	41733
balances	8309	14000
Bank deposits	5390	5076
Cash/bank		
subtotal	20421	113669
Current liabilities		
Creditors	21959	22000
Deferred income	661	850
subtotal	661	22850
Net current assets	77275	90819
Net assets	397347	465038
Loans	-178985	-225000
Total assets	-178985	240038
Financed by		
Share capital	60000	60000
Reserves	10000	13000
Profit & loss	122862	167038
Government grant	25500	0
Shareholders fund	10000	240028

APPENDIX 2: PHASE III RURAL PROJECT (COMPLETED 3/1993)

Profit and loss projections 93/94 - 97/98					
Scenario Includes Incoming Revenue				Amount in 1,000 Dalasis	
Description	93/94	94/95	95/96	96/97	97/98
Rentals	620	648	700	760	818
Local calls	492	566	650	780	930
Trunk calls	5160	5680	6240	6560	6880
Internat. calls	1030	1132	1245	1300	1460
cash national traffic	280	308	368	434	506
Telex/data	30	33	36	40	44
Fees	370	30	55	60	60
subtotal	492	566	650	780	506
Incoming urban	12380	14860	17832	21400	25680
Incoming calls abroad	650	715	786	864	950
Incoming telex abroad	10	10	10	10	10
subtotal	650	715	786	864	26640
Total revenue	**21022**	**23982**	**27922**	**32208**	**37338**
Expenses					
Interest Government	900	900	900	900	900
Staff	800	880	960	1050	1150
Staff (Ops)	600	660	730	800	880
Vehicles	600	660	760	840	1000
Generators	400	440	500	550	600
Maintenance	200	300	400	450	650
Royalties	185	0	0	0	0
Sundries	150	165	180	220	250
Total Expenses	**800**	**880**	**960**	**1050**	**1150**
Profit before depreciation	17187	19977	23492	27398	31908
Depreciation	5000	5012	5036	5060	5100
Net Profit	**5000**	**5012**	**5036**	**5060**	**26803**
Scenario Excluding Incoming Revenue					
Net Profit	**-853**	**-608**	**-136**	**124**	**268**

Telecommunications and Development in Africa
B.A. Kiplagat and M.C.M. Werner, Eds.
IOS Press

CHAPTER XXII

SERVING BOTH MAINSTREAM AND RURAL CUSTOMERS IN BOTSWANA

Mmasekgoa Masire-Mwamba
Group Manager Commercial
Botswana Telecommunications Corporation
Botswana

This article charts the development of Botswana Telecommunications Corporation (BTC) as a typical service provider in a developing country. It considers the challenges facing the corporation and explores how BTC has been able to fulfil its obligations. Although telecommunications service providers in the developing countries are charged with the responsibility of providing adequate telecommunications service throughout the country, they are very often expected to operate on sound commercial basis. It is often an awesome task with the great debate ensuing whether economic development should come before telecommunications development or telecommunications should lead the way. BTC has been able to provide service in the profitable mainstream and the not so profitable rural settings where there is very little economic activity. In recognition of the growing global marketplace, the geographical location, and the advanced local network, BTC has a strategic role in the development of regional telecommunications.

INTRODUCTION

Botswana is a landlocked country in Sub-saharan Africa. It spans an area of 582,000 square kilometres, about the size of France, with only a population of 1.32 million. It is a country of contrasting vegetation with the western part being occupied by the famous Kalahari Desert. Like many African countries, Botswana has a relatively high urban population and a widely distributed settlement in the rural areas.

The urban customer looks to BTC for the most up to date and sophisticated telecommunications service available throughout the world. Business are insisting that the service keeps up with the demands of the emerging information technology. On the other hand, the rural customer merely wants essential service. The rural areas are characterised by sparse population, underdeveloped administration and poor infrastructure and services. These customers want access within a reasonable distance and provision of a service at a reasonable price.

THE CHALLENGE TO BTC

Botswana Telecommunications Corporation was founded in 1980, with two overriding objectives of operating on a sound commercial basis and also of providing telecommunications services throughout the country. BTC was separated from the government department of Post and Telecommunications. A separate Board was appointed whose role it was to ensure that the preset performance criteria were met and that the Corporation met its overriding objectives.

At this stage, BTC inherited a customer base of 6,500 subscriber lines, 0 payphones and 175 telex subscribers. The national economic climate was favorable for Botswana at that time: the highest economic growth periods were experienced in the decade since then, which favoured well for a new corporation with an increasing demand for its services. In 1994, BTC has 40,000 subscriber lines, 500 payphones and 450 telex machines. The line density in Botswana to date is 3.06 extensions per 100 inhabitants, which is well above the Sub-saharan African average.

DIGITAL BACKBONE

How did BTC make these achievements in a relatively short period of time of 13 years that it has been in existence, effectively increasing its customer base almost six fold and also introducing business oriented systems such as Packet Switching and Paging amongst others. Proper fiscal planning and controlled expenditure had to be exercised as we prepared to take on the challenge that had been put before us. BTC embarked on a three phase programme to develop the telecommunications infrastructure of Botswana over digital trunk routes. Phase 1 was completed in 1986 at a cost of over US$ 40 million and covers the South Eastern part of the country. Completion of this phase included the replacement of all electromechanical exchanges which would be served from this microwave link, with digital stored programme controlled exchanges which are called AXE-10. The quoted costs do not include buildings and exchange equipment costs.

Phase 2 was completed in 1991 and covers the Northern part of Botswana. As part of this development, exchange equipment were purchased from Ericsson at a cost of US$ 10 million. This increased the capacity of BTC's switching network by 17 thousand lines. Optical fibre cables were first installed by BTC in 1989.

The third phase is scheduled for completion in 1994 at an estimated cost of US$ 50 million and will cover the South West part of the country, the Okavango and Boteti valley. A trunk route between Jwaneng and Gantsi will complete the digital ring started in phases 1 and 2.

With the completion of the present work, the telephone network will have been extended to cover most of the populated areas of the country. Rural communication is then facilitated by taking advantage of breakout points along the trunk route. The fully digitalised backbone also provides for the connection of telephones and development of other services such as dat transmission which has already been provided, as well as cellular mobile service which is still being considered.

MAINSTREAM CUSTOMERS

Business customers need to be well looked after as they often provide the much needed revenue to bring the appropriate developments to the outer rural communities. Identifying and promoting sophisticated business services is a great challenge as it implies that the dynamics and sophisticated operations of the system be fully understood, usually at great cost. The Botswana Public Switching X.25 network (Botspac) is designed to allow government departments, banks and other users to communicate with the rest of the world, in a way that is cheaper, more reliable and offers more security than regular asynchronous dial-up or leased line communications. Botspac operates over the existing microwave telephone backbone and currently has 19 customers of which the primary user is our national government computer bureau providing access and coordination to several national projects. The introduction of this service has not affected our leased lines and private network business, as it is marketed as an alternative form of data transmission, better suited for particular, specific types of applications.

PROVISION OF RURAL COMMUNICATIONS SERVICES

Rural community settlements tend to be characterised by low population densities spread over a large area. Providing essential services to these customers is of prime importance because very often it is the only form of communication with the outside world. Essential service is defined as the provision of basic telephone service at the

Police station, the administrative centre, the clinic or health facility and the school, as well as a payphone within a convenient distance.

Assessing profitability of providing service in rural areas is very difficult. The costs of providing the service are often higher that in urban areas, the benefits are often non quantifyable as they include social benefits which improve the overall quality of life and encourages local economic activity. BTC took full advantage of its main development trunk route to provide service at breakout points, thereby accessing areas which otherwise would have been very difficult to reach. A commitment has been adopted of providing services to all villages with a population of 500 or more inhabitants. A total of 251 locations has been identified using this criteria and also counting in smaller villages which could be accessed by the breakout points in the Main Development Programme. Of these, 117 have already been provided with varying degrees of service, from payphone to comprehensive service in some cases. Projects under way include the provision of service to 72 additional locations by the end of this current financial year. This would bring the total number of locations to 189 or nearly 75 percent of the large rural settlements. The criteria of 500 residents or more, caters for a large number of villages in Botswana and is a guideline also adopted by government for complementary services.

QUALITY OF SERVICE

In addition to expanding our customer base, it is important that the quality of service be maintained at a high level. This is important not just for the local, but for the international networks as well. Data is compiled on a regular basis to assess the productivity and highlight problem areas. According to the latest compilation of data, BTC's Answer-to-Seizure ratio reflects an average of 46 percent on our international circuits. Although the average is low, we find that for the South African Route we have an ASR of about 60 percent, whereas with the route to Zambia it may be as low as 22 percent. This is an area which needs to be addressed because it means we are not providing the expected service to our customers and therefore we are losing revenue from incompleted calls. We have established that sometimes this is because there is a fault on the route. Whenever the ASR drops on a particular route, investigative measures and corrective action is taken.

IMPACT OF INTERNATIONAL TRENDS ON BTC

The telex service has been negatively affected by international trends in the form of the extended usage of facsimile machines. BTC has however maintained the telex network primarily because of the banks and because its gives better access to some places with a lesser developed telecommunications network. The telex subscriber base has been affected by the takeover of facsimile and has actually reduced its number over the last few years. The number of telex subscribers reached its peak in 1988 at over 700 stations and has come down to 551 in 1993. The convergence of information technology and telecommunications has made it imperative for BTC to provide a fully digital system which can be utilised to provide a better customer service by taking advantage of new technologies.

REGIONAL COOPERATION

The digital ring provided through the phased development programme covers the whole country. It will also provide a fourth international gateway through Namibia, in addition to those through South Africa, Zambia and Zimbabwe, to allow calls to be routed directly from Botswana to the rest of the world. It is a situation which has put Botswana at the forefront of the telecommunications revolution in Africa. An optical fibre route is also being laid alongside the main microwave routes to offer better resilience and higher capacity for traffic handling and processing.

RECOMMENDATIONS

The basis for international cooperation in telecommunications has to be an appreciation and understanding of the current level of development in our individual countries. Only then can the opportunities become clearer and a cooperative process worked out. One area of great concern at a regional level is training. Although there are several regional training centres in southern and Eastern Africa that Botswana subscribes to, there are limitations imposed on the effectiveness of the training programmes offered. The main problem is that they tend to be undercapitalised and therefore cnnot keep up with the technologies and establishment of appropriate training facilities.

Training is complicated by the wide variety in different types of equipment deployed in the region and rapid changes in technology. Administrations therefore invest heavily in training which very soon serves no useful purpose. A regional training requirements assessment would be recommended to identify potential resources for more effective training. Regional training facilities must also be revisited, individual countries could then concentrate on their areas of expertise. One example is that Botswana has already invested in AXE training equipment and could offer regional training on that switching platform, while others concentrate on other aspects.

A consolidation of resources is also very important as regards appropriate research and development targeting local solutions for local problems. One way of achieving this is through the sharing of information on product experiences. Several regional seminars and conferences address this issue and identify common problems. The low level of science and technology development within the region leaves us exposed and vulnerable to technologies which may not be appropriate for our applications. Priority needs to be attached to this aspect of cooperation.

FUTURE DEVELOPMENTS

Future developments in BTC include liberalisation of customer apparatus. This is a way to increase the participation of terminal equipment providers into the local telecommunications market. It provides our mainstream customers the choice of system and functionality which can only be made available through a wider supplier base. Furthermore, it gives BTC the opportunity to concentrate on network provision. This liberalisation programme is scheduled for 1994/1995.

Value added services are a way of increasing the usage of our network. There are several options available, but an evaluation of these is necessary before adoption can be decided upon. Provision of payphones with an emphasis on card identification and payment, both for urban and rural applications, is being considered for introduction on a national scale.

Rural communications remains a priority for BTC and all efforts are being made to address rural communities' needs alongside our mainstream customers.

Table 1 : Statistics 4th quarter 1993

New Applications	1,594
Waiting List	12,283
Ceased Lines	743
New Lines Connected	1,775

Table 2 : Summary of Statistics (Totals)

Exchange Lines	41,599
Stations	73,019
Private Wires	457
PABX	2,094
Exchange Capacity	56,289
Pagers	258
Payphones	513
Packet Switching	19
Telex	394

Table 3 : Forecast of Telephone Demand 1993 - 1998 (at 31st March)

Exchange Area	1993	1994	1995	1996	1997	1998
Gaborone	24,544	27,508	31,643	36,370	40,745	44,819
Lobatse	4,854	5,422	6,235	7,17	8,246	9,482
Palapye	4,68	5,238	6,023	6,027	7,966	9,161
Southern	34,078	38,168	43,892	50,476	56,957	63,462
Francistown	11,414	12,756	14,670	16,870	19,399	22,310
Selibi-Phikwe	3,213	3,433	3,948	4,540	5,221	6,004
Northern	14,627	16,189	18,618	21,140	24,620	28,314
Total	48,705	54,357	62,510	71,886	81,577	91,776

Graph 1

Forecast of Telephone Demand 1993-1998

(as at 31st March)

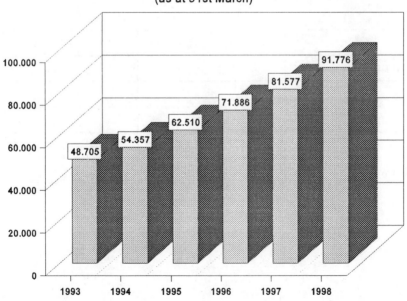

Graph 2

Working Lines as at 31 March

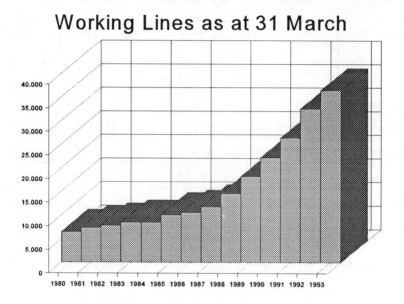

Table 4 : Quality of Service Incoming International
(Answer to Seize Ratio quarterly averages, measured locally), in %

	previous quarter	last quarter
United Kingdom	35.6	50.5
USA	50,6	450
South Africa	57.0	45.0
Zambia	22.8	38.5
Zimbabwe	45.1	51.4

Telecommunications and Development in Africa
B.A. Kiplagat and M.C.M. Werner, Eds.
IOS Press

CHAPTER XXIII

FINANCING THE DEVELOPMENT OF RURAL TELECOMMUNICATIONS : THE KENYAN EXPERIENCE

Rufus Muigua
Assistant General manager Telecommunications Services
Kenya Posts & Telecommunications Corporation
Kenya

ABSTRACT

A survey of the distribution of telecommunication facilities in any country today would reveal substantial imbalances between the rural and urban areas. Yet it is recognized that the key to stimulating the development of productive enterprise in the rural areas is the provision of basic infrastructures including telecommunications. The paper looks at strategies that Kenya has adopted in its efforts to redress the existing imbalances and further examines options for financing rural telecommunications development.

BACKGROUND

POLICY FRAMEWORK

The key to stimulating the development of productive enterprises in the rural areas, which provide residence to the majority of the population and have vast untapped potential, is the provision of requisite basic infrastructures and enhancement of the linkages, or webs of interchange, that exist between rural and urban areas. In recognition of this fact, therefore, Kenya's approach to rural development has been based on the "growth centre" strategy whose major aim is to provide basic infrastructural facilities at designated urban, rural and market centres which will serve as focal points for the development of rural areas. The role of telecommunications has been explicitly recognized within the overall framework of the country's integrated rural development approach. The expansion and modernization of rural telecommunication facilities has, therefore, been a major policy objective in recent years.

STATEMENT OF THE PROBLEM

The country has made encouraging progress in its efforts to expand and modernize rural telecommunications facilities. However, it is recognized that substantial differences still exist between urban and rural areas regarding access to tele-communications facilities. A major constraint in this regard has been inadequate financial resources. Rural telecommunications require relatively large capital investments and have relatively long payback periods, which necessitate long-term financing arrangements. Bilateral and multi-lateral development assistance has largely targeted other infrastructural facilities such as roads, water and power supply.

This paper examines the progress being made towards redressing the rural-urban disparities regarding access to telecommunication services in Kenya and options available for the financing of rural telecommunications.

STRATEGY FOR THE DEVELOPMENT OF RURAL TELECOMMUNICATIONS

GROWTH CENTRES STRATEGY

In line with the Government's integrated rural development strategy, telecommunication services have been provided at designated growth centres since the Plan period 1979-1983. The service provided, depending on the level of demand, was either a manual exchange or a public payphone or both. However, from 1984, the policy emphasis shifted from mere extension to expansion and modernization of rural telecommunications. In this regard, the Rural Automation Project was formulated and resources mobilized to facilitate its implementation. The project was to be implemented in 3 phases. Phase I covered all 51 district headquarters. Phase II covered 314 divisional headquarters and high revenue generating areas. The 3rd phase was to cover the remaining centres within the framework of the growth centre strategy. To date phase I has been wholly implemented and the implementation of phase II is in progress. The distribution of these centres is presented in Annex I.

CHOICE OF TECHNOLOGY

As indicated above, the initial intervention in the rural areas was mainly in the form of manual telephone exchanges and radio call services in the remote areas. These were supported by rural carrier systems and single channel radios. However, with the move toward modernization, KP&TC adopted UXD-5B digital exchanges of 300 lines and 600 lines capacity. In addition, line concentrators, and digital multi-access radio systems have been successfully installed to provide, cost effectively, modern telecommunication services to pockets of subscribers in rural and remote areas as well as lodges in Game parks. The choice of this technology was arrived at after extensive evaluation of technological options available in the market.

Possible application of radio-based subscriber loops and cellular mobile is being investigated, though these might initially target a specialised clientele.

It is however noted that even with the application of these modern innovative technologies, the cost per line of providing service in the rural areas remains considerably higher (ratio 2:1) than in the urban areas.

RESOURCE MOBILIZATION

Kenya's policy with respect to financing telecommunications development programmes has always been to borrow externally in order to cover foreign exchange costs. Retained earnings have been used to finance local costs. This policy is justified on two grounds. First, the country has not developed a capability to manufacture telecommunications equipment to meet local needs. Any development of the telecommunication network, therefore, must necessarily involve foreign exchange expenditure. Secondly, the foreign exchange constraints that the country, and indeed most developind countries, experience have left Kenya with no alternative but to seek external sources of finance to facilitate implementation of telecommunications development programmes which on average, consist of a 70-90 per cent foreign exchange component.

Since 1984, Kenya has invested approximately US$ 270 million in the expansion and modernization of rural telecommunications. 90 per cent of this amount (or US$ 246 million) was in the form of loans and the rest was internally generated. it is noteworthy that bilateral and multi-lateral development assistance accounted for only 15 per cent (or US$ 37.4 million) of the debt financing, thus confirming the reluctance of these institutions to finance rural telecommunications. Most of the financing was commercial loans in the form of suppliers' or export credit.

STATUS OF THE TELECOMMUNICATIONS NETWORK

GROWTH OF THE NETWORK

It is noted in Annex II (a) that over the reference period (1981-1993) total telephone exchange capacity increased at an annual growth rate of 10 per cent rising from 112,861 lines in 1981 to 361,840 lines in 1993. The rural network expanded at the rate of 11 per cent per annum whereas its component in the exchange capacity increased from 16.6 per cent in 1981 to 24.3 per cent in 1990 before declining to 19.1 per cent in 1993. Exchange connections in the rural areas increased at an annual growth rate of 10 per cent as compared to the 8 per cent growth rate registered by the urban network. Inadequate external line plan has been identified as the major constraint in the expansion of exchange connection particularly, in the rural areas.

PUBLIC TELEPHONES

The expansion of the public telephone (coinphone) service was the most outstanding achievement in the reference period. The number of telephone booths increased from 588 in 1991 to 5,671 in 1993, thus registering a growth rate of 21 percent per annum. The expansion programme which covered the urban and rural areas was financed by a grant.

LEVEL OF AUTOMATION

The on-going implementation of the Rural Automation Project has already had a major impact on the quality of telecommunication services in the rural areas. As indicated in Annex II (b) the proportion of automatic (STD) exchange connections (ECs) increased from 15.3 per cent to 68.3 per cent during the same period.

SERVICE PENETRATION

Service penetration in the rural areas doubled from 0.07 to 0.16 ELs per 100 population during the reference period. However, despite the substantial investments already committeed towards the expansion and modernization of rural telecom-munication the level of penetration is still low compared to the urban and rural areas regarding access to telecommunications facilities.

CAPACITY UTILIZATION AND REVENUE GENERATION

CAPACITY UTILIZATION

Capacity utilization in the rural areas, indicated by the exchange fill statistics in Annex II (a), has ranged between 46.1 per cent and 54.9 per cent during the period under review. This compares poorly with the urban areas where capacity utilization, during the same period, varied between 61.7 per cent and 81.8 per cent. However, an analysis of sampled exchanges depicted in Annex III and IV shows clearly that rural areas are not homogeneous. In the case of district headquarters (Annex III) for instance, it is noted that capacity utilization, particularly after automation, has been relatively high and, in fact, the 1993 figures are higher than the urban average of 61.7 per cent recorded in 1993. The divisional headquarters (Annex IV) on the other hand, indicate a lower capacity utilization as compared to district headquarters and the urban average. The existence of fairly significant waiting lists in all cases points to inherent constraints in connecting new subscribers.

REVENUE GENERATION

Annex V and VI analyze revenue generation by a sample of exchanges at the district and divisional headquarters in the period just before and after automation. It is noted that in virtually all cases revenue doubled in the period after automation. There is also a significant rise in revenue per line in the period after automation. The district headquarters, invariably, generate higher revenues than the divisional headquarters.

The above analysis, whose timeframe was deliberately narrowed to preclude the influence of tariff changes, tends to imply that appropriate new technologies could substantially improve revenue generation capacity of rural telecommunications. The analysis also exposes the fallacy of perceiving rural population as incapable of appreciating modern technology.

OPTIONS FOR FINANCING RURAL TELECOMMUNICATIONS

EXTERNAL FINANCING

External financing in the form of loans and grants will cotinue to play a critical role in the development of rural telecommunications. The Kenyan experience has indicated the growing contribution of commercial sources such as banks, suppliers and export credit agencies in financing telecommunications development. Arising from this experience, however, a number of critical issues have emerged which need to be addressed when formulating an appropriate borrowing strategy for rural telecommunication development.

❐   A comprehensive financial plan setting out the capital costs for infrastructural development with realistic financial projections of income and operating costs is an essential planning tool for sustaineable rural telecommunications development.

❐   The determination of level and type of external debt that is manageable in the short run and sustainable in the long run. This must be done in the context of corporate financial position and the country's balance of payments status.

❐   The choice of currencies, exchange rate fluctuations, maturity structure, grace period and balance between fixed and floating rates of interest are crucial considerations in the search for the least cost forms of borrowing.

INTERNAL CASH GENERATION

Internal cash generation is the most reliable source of finance from the point of view of telecommunications administrations. Its significance as a source of funds for telecommunication investment is bound to grow in a resource constrained environment. However its scope for further expansion is dependent on the following factors:

❐   The formulation and implementation of tariff strategies that adequately reflect the cost of service provision and also encourage efficient utilization of the network;

❐   Improvement in quality of service particularly call completion rates and fault clearance which account for a substantial proportion of lost revenue.

❐   Improvement in billing and debt collection. Billing disputes in particular tend to tie up much needed revenue;

❐   Hedging against or limiting exposure to foreign exchange losses which tend to erode earnings which would otherwise be retained for development purposes.

RURAL TELECOMMUNICATIONS DEVELOPMENT FUND

A special fund, to be administered by the Government or its designated agent, could be set up specifically for the purpose of developing rural telecommunications. Under this arrangement, contributors would include the Government, local authorities and donor agencies. Telecommunication administrations would also be expected to contribute a percentage of their earnings towards this fund. The management of the fund would be modelled along the lines of similar funds such as the Rural Development Fund and the Rural Electrification Fund. Consequently, the development of rural telecommunications will be integrated within the overall national plans for rural development.

TELECOMMUNICATION TAX

A telecommunication tax could be an important source of funds for the development of rural telecommunications. Under this arrangement, all subscribers would be expected to pay a certain percentage of their monthly bills as tax for purposes of network development. The rationale, in this regard, is that the benefits of expansion of the network accrue to all subscribers by virtue of the increase of calling opportunities. Indeed it has been estimated that the benefit of expanded calling opportunities increases at a rate more than twice as fast as the network's expansion rate. The challenge, however, would be in convincing the government, which is faced with competing demands for limited resources, to avail telecommunication tax revenue for the exclusive purpose of developing the rural telecommunication network.

TELECOMMUNICATION DEVELOPMENT BONDS

Under the scheme of telecommunication development bonds subscribers who seek telephone service would be required to purchase a minimum amount of bonds which would mature after a specified period and accrue attractive interest. This mode of financing telecommunication development has been successfully utilised in other countries, e.g. Japan.

CONCLUSION

The need for a balanced approach to the development of telecommunications services where appropriate information and communication technologies are introduced to the rural and remote areas no longer need emphasis. The need for telecommunication intervention in the context of diversifying the rural economy and creating the infrastructure for new industrial and commercial initiatives is now clear. Recent technological advances in information and communications present unique opportunities for the development of rural areas. What is required is the identification of least cost methods of financing and the mobilization of all the actors, government, private sector, financial institutions and development assistance agencies, to collaborate in the development of rural telecommunications.

Manufacturers and international standardisation bodies must also endeavour to come up with more innovative and low cost technology that will make the telephone affordable to the ordinary person in the rural areas.

OoOoOoO

RURAL TELECOMMUNICATION DEVELOPMENT

Comparision of Rural and Urban Networks

YEAR	1981	1987	1993	AVERAGE ANNUAL GROWTH %
EXCHANGE CAPACITY				
(a) Urban	94,076	174,750	292,648	10.0
(b)Rural	18,785	44,330	69,200	11.0
(c)Total	112,861	191,750	361,848	10.0
Rural Component in Total Exchange Capacity (%)	16.6	23.1	19.1	
EXCHANGE CONNECTION				
(a) Urban	68,635	136,602	180,426	8.0
(b) Rural	10,229	22,197	32,770	10.0
(c)Total	78,864	158,799	213,196	9.0
Rural Component in Total ECs (%)	13.8	14.0	15.4	
PUBLIC TELEPHONES				
(a) Urban	302	2,530	3,913	23.0
(b) Rural	256	1,041	1,758	17.0
(c) Total	588	3,571	5,671	21.0
Percentage of Rural Public Phones	43.5	29.2	31.0	
EXCHANGE FILL (%)				
(a) Urban	73.0	78.2	61.7	
(b) Rural	54.5	50.1	47.4	

RURAL TELECOMMUNICATION DEVELOPMENT
Automation and Penetration Levels

YEAR	1981	1987	1993	AVERAGE ANNUAL GROWTH %
RURAL EXCHANGE CAPACITY				
(a) Automatic	2,170	17,000	44,800	28.0
(b) Manual	16,615	27,330	24,400	30.0
(c) Total	18,785	44,330	69,200	11.0
Level of Automations (%)	11.6	38.3	64.7	
RURAL EXCHANGE CONNECTIONS				
(a)Automatic	1,569	8,668	22,387	25.0
(b)Manual	8,660	13,529	10,383	1.0
(c)Total	10,229	22,197	32,770	10.0
Level of Automations (%)	15.3	39.1	68.3	-
Total Population (In Millions)	17.3	21.6	26.2	3.4
Rural Population (In Millions)	14.7	18.2	21.0	3.0
Rural Penetration (DELs per 100 Population)	0.07	0.12	0.16	-
National Penetration (DELs per 100 Population)	0.46	0.73	0.81	-

NOTES

1. Municipalities with commercial power supply have been classified as Urban.
2. Municipalities without commercial power supply and all other centres have been taken Rural.
3. DELs = Direct Exchange Line.

EXCHANGE GROWTH BEFORE AND AFTER AUTOMATION
(The case of District Headquarters)

NAME OF EXCHANGE	YEAR	1984	1986	1988	1993
	PARTICULARS				
KILIFI	Capacity	300	300	600	600
	Connections	194	209	336	441
	Waiters	64	8	58	63
	Exchange Fill %	65.3	69.7	56.0	73.5
NAROK	Capacity	140	140	600	600
	Connections	125	129	238	436
	Waiters	79	108	29	105
	Exch. Fill %L %	89.3	92.1	39.7	72.7
SIAYA	Capacity	300	300	600	600
	Connections	130	133	242	375
	Waiters	116	172	147	222
	Exchange Fill %	43.3	44.3	40.3	62.5
BUSIA	Capacity	300	300	600	600
	Connections	141	164	238	480
	Waiters	61	53	81	86
	Exchange Fill %	47.0	54.6	39.6	80

NOTE

First 3 years (1984-86) denote period before automation while the period (1987-1993) is after automation.

EXCHANGE GROWTH BEFORE AND AFTER AUTOMATION
(The case of Divisional Headquarters)

NAME OF EXCHANGE	YEAR / PARTICULARS	1986	1988	1990	1993
SAGANA	Capacity	140	140	300	300
	Connections	98	106	137	209
	Waiters	106	113	161	187
	Exchange Fill %	66.4	75.7	45.7	69.2
SOTIK	Capacity	140	140	600	600
	Connections	94	120	176	185
	Waiters	145	134	194	103
	Exchange Fill %L %	67.1	85.7	29.3	30.8
MAGADI	Capacity	140	140	300	300
	Connections	71	93	92	106
	Waiters	27	39	59	30
	Exchange Fill %	50.7	66.4	30.7	35.3
TAVETA	Capacity	140	140	300	300
	Connections	71	106	167	207
	Waiters	30	67	72	60
	Exchange Fill %	50.7	75.7	55.7	69

NOTE

First 3 years (1986-88) denote period before automation and the period (1989-1993) is after automation.

REVENUE GENERATION BEFORE AND AFTER AUTOMATION
(The case of District Headquarters)

PERIOD	SIX MONTHS BEFORE AUTOMATION		SIX MONTHS AFTER AUTOMATION	
EXCHANGE NAME	Half-year Revenue	Revenue per* Line	Half-year Revenue	Revenue Per* Line
KILIFI	19 149.40	94.80	39 001.20	128.70
NAROK	11 837.90	91.80	25 625.90	105.90
SIAYA	12 342.30	92.80	30 223.50	154.20
BUSIA	15 914.10	97.00	36 693.80	163.80

1 US Dollar = KShs 68.00

NOTE
* Revenue per line is calculated for the half-year period

REVENUE GENERATION BEFORE AND AFTER AUTOMATION
(The case of Divisional Headquarters)

PERIOD	SIX MONTHS BEFORE AUTOMATION		SIX MONTHS AFTER AUTOMATION	
EXCHANGE NAME	Half-year Revenue	Revenue per* Line	Half-year Revenue	Revenue per* Line
SAGANA	5 153.20	48.60	5 258.70	42.40
SOTIK	10 400.00	86.70	15 205.00	95.60
MAGADI	10 172.60	109.40	16 881.30	205.90
TAVETA	8 182.10	77.20	17 732.10	127.60

1 US Dollar = KShs 68.00

NOTE
* Revenue per line is calculated for the half-year period

Telecommunications and Development in Africa
B.A. Kiplagat and M.C.M. Werner, Eds.
IOS Press

CHAPTER XXIV

TANDEM COOPERATION BETWEEN NEIGHBOURING COUNTRIES : GUINEA - SIERRA LEONE - LIBERIA

Ernest Olumide Blyden Johnston
Deputy General Manager
Sierra Leone External Telecommunications Ltd
Sierra Leone

It is depressing to note that at the time of preparing this paper, none of the telecommunication links between the three neighbouring countries Guinea, Sierra Leone, Liberia worked. This paper therefore concentrates on some of the problems encountered with the Panaftel links, whose objectives were to provide a good telecommunications infrastructure within the participating countries while at the same time providing cross border communication. Much can be achieved with efficient cross border telecommunications, since it serves common interests of the entire society and its economy.

However, judging from experience with Panaftel, it appears that telecommunications provision is not always customer driven but rather left to major telecommunication operators to develop and manage. Traffic between neighbouring countries remains minimal because of the high costs involved, because of using transit routes often through Europe or North America. Traffic is routed automatically through such transit lines, since Panaftel is not operational. This paper will not apportion blame to any party, but rather narrate collective strategies agreed upon which have not come into fruition. The West African economic cooperation organisation ECOWAS has sent technical delegations to those countries which experienced difficulties in making Panaftel operational, however, these mission have not yet produced the envisaged results.

The Panaftel links currently under discussion in Guinea, Sierra Leone and Liberia are the following:

1. Route Path Sierra Leone - Guinea

⇨   Freetown - Port Loko - Kambia - Conakry - Wonkifong - Coyah - Forecariah

2. Route Path Sierra Leone - Liberia

⇨   Freetown - Rotifunk - Moyamba - Mano - Bo - Bandajuma - Monrovia - Mano Mines

As the history of Panaftel shows, strong commitment to regional cooperation is a prerequisite for an efficient and reliable joint network. This point is illustrated by the problems which were incurred in guaranteeing power supply along the routes. During the 1980's, several of the reported problems of the network were attributed to logistical support for electric power at remote stations along the Panaftel routes. As lessons were learnt, the problem of securing power is becoming alleviated since the tradition method of power supply was replaced with solar systems, providing the much needed additional reliability for the long haul links. Sierra Leone has been able to install solar power in all the repeater stations outside the bigger towns and the links in between them are now functioning satisfactorily. Meanwhile however, these installations require permanent surveillance or protection, which again is a matter of commitment on the part of participating telecommunications operators.

The switching capacity has been dimensioned to take on cross border traffic, but for reasons which will be highlighted later, these investments have not been utilised.

Guinea, Sierra Leone and Liberia are Members of the Mano River Union and also Members of ECOWAS. Efforts have been made by both organisations in arranging coordination meetings and providing technical support to get the regional cross border links operational.

The problems associated with getting the links to function could be attributed to the following causes:

❖ Technical identification of the problems;
❖ Insufficient commitment of technical staff;

❖ Prioritising of cross border links at senior management level;
❖ Lack of technical expertise in transmission and interfacing of networks;
❖ Financial gains of cross border links for operating companies not sufficiently recognised.

Many of these problems have their roots in general conditions of the structuring, management and regulation of the economies in a number of African countries. In the 1980s, following independence, African Governments were advised to participate in business concerns in their countries. As a result, many parastatals were created and later, shares were sometimes floated. In most cases however, governments were the sole shareholders whereas in the many other instances they were majority shareholders. Only in very few cases were they the minority shareholders. The impact of these parastatals on society and economy was negative. Politicians overstaffed these organisations in their quest to find jobs for "the boys". Their nominees were often incompetent, inefficient and arrogant. The parastatals therefore were managed along unprofitable lines and the respective governments levied taxes on the population in order to find funds to continue running their operations. It was not uncommon that a Board of Directors was appointed at the whims and caprices of politicians. As their protegees, those directors were rather loyal to them than to the parastatals for which they carried responsibility.

After decades of decadence, the negative effects of this situation was finally diagnosed and the view that governments should undo what they had created gained ground rapidly. More and more citizens in African countries agree that their governments should never be involved in commercial ventures, since accountability is difficult to come by. In parastatals there is not enough incentive and lack of motivation, as management realised that losses were absorbed by the levying of more tax burdens on society.

In Sierra Leone, the Government has set up a Public Enterprise Reform and Divestiture Commission (PERDIC) to privatise most of the parastatals and restructure those that are considered economically viable. The econbomic climate being created favors accountability and transparency. The Government has finally accepted that its major role is to create an atmosphre of confidence in the business sector which will result in gaining more substantial revenue from taxes and license fees. The public sector's performance in the economy can now be considered as only minimal. For example, previously road building and maintenance was being undertaken by the public sector, the Ministry of Works. The establishment had nearly 20,000 labourers to work on the road network. However, the state of the roads was deplorable. This labor force has since been made redundant and a new Sierra Leone Roads Authority was set up with less than 1,000 labourers, whose performance is highly commendable and whose presence is felt. Foothpaths have been repaired, roads resurfaced and bridges that had disappeared have been rebuilt. In the same vein, because of liberalised trade, healthy competition by way of advertisements in the newspapers as well as on the radio has helped the 'fourth estate', the press, to function better. In the recent past, newspapers such as the Daily Mail used to be marketed sometimes only once a month. The radio is now on air at regular times daily.

In the Acts prior to 1992, provision was made for an aggrieved party top appeal to the concerned Minister. In legal practice of today, this is being eliminated and an aggrieved party would have to appeal to the Courts. The role of the Judiciary is strengthened. In summary, public sector involvement in the economy is being phased out and Government is creating an atmosphere where the private sector could invest without seeking permission for a fee from a politician. This is eliminating corruption. or at least, bringing it down to a minimum. Today, investors need only purchase the Policy Guideline on the area they wish to invest in and find all necessary regulatory details.

Initially, Panaftel was designed to be an analogue network, linking almost all African countries. From the beginning, Panaftel constituted an important case for countries to establish cooperation in telecommunications at a most practical level. However, some of the Panaftel links are beginning to outlive their usefulness, partly in view of the digitalisation of telecommunications networks which is a dominating trend, also in Africa. It is therefore recommended that neighbouring countries agree to coordinate their network planning, to ensure that Panaftel can succesfully upgrade its links from analogue to digital. After all, Panaftel has an important role to play, next to the national operators and next to the African satellite organisation RASCOM.

OoOoOoO

APPENDIX

PANAFTEL AND RASCOM - BACKGROUND (*)

The idea for a Pan African telecommunications network PANAFTEL was born in 1962, at the first meeting in Dakar (Senegal) of an ITU Plan Committee which was assigned to elaborate a general plan for telecommunications in Africa. PANAFTEL was originally designed to be the backbone of a regional network, formed by the interconnection of national networks with each other and with other continents. A Coordinating Committee composed of representatives of the OAU, ADB, ECA, ITU and PATU set out to supervise and guide the project. Subregional coordinating bodies UAPT, ECOWAS, UDEAC, SADCC, WADB, BDEAC and ABEDA became also involved in implementing the network.

Implementation of PANAFTEL only began in 1974. As of now, PANAFTEL radio-relay systems on routes totalling some 40,000 km in length have either been or are in the process of being installed. To complement these transmission routes, 39 inter-national telephone switching centres have also been or are in the process of being installed. Moreover, nearly 8,000 km of submarine cables have been installed and 42 out of the 45 countries participating in the project have earth stations for staellite communications. Financing for remaining missing links has to be found to complete the basic PANAFTEL network. At present, availability of the network is not ensured all over. Efforts should becontinued to improve operational efficiency and attain tariff harmonisation. The physical installation of the main links in the West African subregion has been completed, but operational efficiency and harmonisation of tariff structures need to be greatly enhanced.

Part of the task of providing connectivity for the African countries is also being assumed by RASCOM (Regional African Satellite Communication System). The initiative to RASCOM dates back to 1976, when the Conference of African Ministers of Transport, Communications and Planning held in Port Louis, Mauritius, decided to undertake a feasibility study. The Lead Agency for the feasibility study was the ITU. There is broad consensus among all governments to press ahead with the initiative.

RASCOM had entered its operational phase towards end of 1993 and it working as an autonomous operating entity. Pooling of existing satellite space segments contracted by African countries is currently in progress. The first RASCOM transponder was contracted as per 1 February 1994 by a South African television broadcast group. RASCOM is negotiating and/or preparing coordination with PANAFTEL, with submarine cable initiatives targeted at Africa, as well as with Low Earth Orbit (LEO) satellite telecommunications projects.

(*) Source : ITU.

CHAPTER XXV

PRIVATIZATION OF THE TELECOMMUNICATION SECTOR IN THE SUDAN (*)

Dr. Idriss Ahmed Yousif
Deputy Director General
Sudan Telecommunications Public Corporation
Sudan

INTRODUCTION: THE CHALLENGE OF A VERY LOW PENETRATION

The telecommunication penetration in Sudan stands around 0.3 telephones per 100 inhabitants which is one of the lowest of all countries in the world. Translated into installed equipment the network is comprised of the following components:

- ❖ 75,000 telephone lines
- ❖ 2,300 kilometers of microwave links and microwave link to Saudi Arabia
- ❖ 1 international earth station with 33 SCPC channels (Intelsat)

❖ 1 regional earth station (Arabsat) with over 120 channels to the various Arab countries.

❖ 15 earth stations forming the domestic satellite network

❖ 1 international telephone exchange

❖ 1 international telex exchange

The Government in its current overall 10 year National Strategy for Development has anticipated an expansion in the telecommunication sector from 0.3 main lines per 100 inhabitants to 2 lines per 100 inhabitants. This means that the present telephone density has to be increased to more than 9 times in the coming ten years. This should boost the number telephone lines from 75,000 lines to 650,000 lines.

The minimum investment required to achieve this goal will be over US$ 1 billion. Certainly, the Government with its various financial constraints and responsibilities, especially in the first priority sectors such as agriculture, defense, security, health and education, will not be able to contribute substantially to the development of the telecommunications sector. This has been the main reason why this sector has not progressed over the past few years. In addition to the fact that whatever revenue has been earned by this sector, it has been used in other urgent immediate needs rather than investing it in the telecommunication sector.

As a result of this lack of proper development in telecommunications, the private sector resorted to very expensive isolated and therefore inefficient private solutions. Examples are low-capacity HF and VHF links, and even sometimes resorting to the very expensive Inmarsat services using small earth stations.

It is clear that if the national strategic target of 2 telephones per 100 inhabitants in the coming ten years is to be achieved, together with a substantial improvement in the quality of services, then the private sector which is the major user of telecommunications, has to be involved one way or the other, and the means and mechanisms have to be established to encourage the private sector to invest in this sector.

The Government, as a result of a series of conferences on National Economical Salvation Policy, has adopted the policy of privatisation in various sectors, especially those of commercial nature which were run by the Government and in the field of profitable projects as those which need heavy foreign currency investments. The telecommunication sector has been one of the targeted sectors in this policy.

The telecommunication sector was identified as one of the important sectors to be privatised, not for being unprofitable but for not being able to meet the national requirements and for not being able to raise or obtain funds or investments for the very badly needed expansion and development projects.

Sudan Telecommunication Public Corporation (STPC) has been the sole body for providing public telecommunication services and establishing the networks for public communications, relaying sound and television programmes to the broadcasting stations and in most cases planning, operating and maintaining these services and other government private networks. STPC was also undertaking all the regulatory functions, especially those of licensing.

A Higher Level Committee was therefore formed to look into the practical steps of achieving these policy goals in an orderly manner.

GOVERNMENT APPROACH TO PRIVATISATION

The first steps taken towards privatisation started when the private sector was allowed to import telecommunication terminal equipment such as telephone sets, telex machines, fax machines, PABX's, as well as secondary and sometimes primary telephone cables. The breaking up of the monopoly of STPC in this field was for import for personal use or for commercial purposes, all in accordance with stict laws and regulations established by STPC and authorized by the Council of Ministers.

The next positive step towards privatisation was the formation of a High Level Ministerial Committee for privatisation of various sectors, of which telecommunications was one. This Committee formed a Technical Committee for preparing a report on how privatisation of the telecommunication sector could be achieved.

The Technical Committee formed a number of sub-committees to carry out its mandate. The following sub-committees were formed:

- ❖ Committee for the evaluation of STPC assets;
- ❖ Committee for financial issues and debts, and legal issues;
- ❖ Committee for presenting structures, functions, duties and authorities of bodies resulting from the breaking up of the functions of STPC and the procedure for implementation of privatisation;
- ❖ Committee for personnel aspects;
- ❖ Committee for building plans.

The above sub-committees were further reshuffled into a number of sub-subcommittees which produced detailed reports on their terms of references.

The Committee on evaluation of assets was confronted by the issue of choosing between the major evaluation methods normally used. These were: I) the replacement-method of evaluation; II) the current market-price method; and III) the revenue evaluation method.

Because equipment was a mixture of old and new equipment and because those that had spent more than their expected life time were still working and providing services, it was decided that a replacement method of evaluation be used and the present technical status and the expected remaining life time be estimated hence its replacement value be obtained. This has been the method used for the initial evaluation undertaken. However, at the stage of forming the new public company, the consultants recruited for re-evaluation used the revenue method which is basically an evaluation for the business based on traffic forecasts and equipment performance. The revenue method of evaluation therefore does not go to the level of specific assets or listing assets in an itemized manner.

The Committee on financial and legal issues has worked on debts to other entities or foreign administrations and the telephone debts requested from the subscribers and the short-and long-term loans obtained by the Government which are still to be paid, the money to be put aside for compensation to the personnel who are to leave their jobs as a result of privatisation. The issues of separating accounts local or international as a result of having two telecommunications service providers had to be addressed as well.

The Committee on restructuring and proposals for the transitional period has formulated proposals for the transitional and the new structure of the remaining STPC and has also proposed the towns which are to be part of the new company. The selection comprises all the important commercial and capital centers from various states including the capitals and the microwave links, with the anticipation that in the future other important cities can be added to the new company. Under the above arrangement, only rural areas will be left to STPC to manage and develop. The Committee also proposed the formation of a Regulatory Body to undertake the functions of sovereign nature such as licensing, standardizing, approving tariffs, spectrum management, representation of the state in international organizations and conferences.

The Committee on restructuring also proposed that the other functions which were carried by STPC in the field of planning, operating and maintaining sound and TV broadcasting stations be immediately transferred to the Sound and TV Broadcasting Organizations. The relay of sound or TV programmes to the broadcasting sites could be undertaken by either the new company or STPC, depending on the type of activity existing in each one.

The personnel Committee has listed all personnel presently working with STPC and has calculated their compensations in the event that they leave STPC or are separated by the new company. The Committee is also aware of the total amount of money to be made available for this purpose.

The final report of the Technical Committee which proposed the establishment of a shareholders' public company in which the Government will participate by its assets and the formation of a Regulatory Body to supervise and regulate the telecommunication activities in the new era after privatisation of SPTC. The report outlined the demarkation lines between STPC and the new company. The Committee further proposed that two new committees be formed, one to outline the practical steps for forming the new company, and the other to submit proposals on the detailed functions and authorities and structure of the Regulatory Body. These two committees were formed, the results of which were the formation of the public shareholder' company and the National Telecommunication Council, as discussed below.

288

ESTABLISHMENT OF SUDATEL

The promotion committee for establishing the public shareholders' company prepared all the necessary promotion documents such as deciding on the capital of the company in foreign currency and in local currency and the value of each share.

In order to speed up the arrangement, a group of promoters who are also potential contributors to the new company, was formed. It was decided that the deadline for the formation of the company be set for 15/11/1992; in addition,the founder-shareholders were to deposit their shares which were a minimum of US$ 100,000 per founder-share to an account opened for that purpose and prior to 15/11/92.

Concurrently, an independent appraiser was contracted to evaluate the assets of STPC with which the Government will contribute in the new company. In addition, draft shareholders agreements to establish the company and draft agreements between the company and the Government were prepared for discussion and approval.

The Government took very drastic measures in encouraging the private sector to participate in the proposed new company. These measures can be summarized as follows:

❐   The Government has been conservative in the value of assets it is to contribute in this company to encourage and ensure that the private sector participate;

❐   The Government has reduced its supervisory role in the new company so that its representation in the Board of Directors does not exceed 20% of Board members, even if the contribution of the Government is more than 20%, which is to be reduced gradually through selling of its shares to the public so that the performance of the company is not affected by Government procedures. Shareholders are both national and foreign.

❐   The Company was given as an initial step concession to provide srevice to all those towns which are commercially viable, including as a whole of the capital and the major social and commercial centers;

❐   The Company has been given a monopoly or lease of 15 years of international telecommunication services, renewable every 5 years.

❐   The supervision of the Minister of Communication and Tourism or the Government in general has been limited to its representatives in the Board of Directors;

❐   The new company has been given all the incentives and privileges given to the local and foreign investors by the 1925 Act and the more recent acts passed by the   Government such as the 1990 Investment Encouragement Act and exemption from import duties and certain taxation fees;

❐   The company is also given a free hand to choose from the present skilled personnel of STPC or from abroad. In fact, it has relied on foreign expertise in its top level management, to be assisted by skilled personnel from STPC. An agreement was signed between Sudatel and Nepostel of the Netherlands to provide temporary management.

In spite of the very many important incentives given to the company, shareholders' initial contribution did not meet expectations. Nevertheless it is anticipated that once the company is fully in control, the shares will be open to the public (25% of the capital); it is then expected that the company will attract more capital, and communication will certainly start to develop and expand with the positive contributions of shareholders and the lending banks and the shift to the commercial operation instead of the previously Government-run Corporation which was constrained by the slow Government administrative and financial procedures.

The process of transferring the assets of STPC that belong to the company as a contribution of the Government in this company was effected on 13/9/1993. A transitional period of four months was agreed upon in order to achieve the following goals:

❐   separation of accounts between the new company and STPC since the two will be working side by side through reading of meters and finalizing of debts with former STPC subscribers;

❐   sorting out of foreign debts;

❐   introducing the new company to the foreign administrations and international organizations;

❐   division of personnel between the company, STPC and the Regulatory Body in addition to the Sound and Broadcasting Organizations;

❐   formulation of traffic agreements and rates for traffic flowing between STPC and the company networks whether nationally or internationally.

The transitional period was a very difficult one for both personnel, the company and STPC in surpassing the stage of not belonging to any and not knowing what will happen next. The new company has been in total control of its areas of activity and has decided on its personnel since February 1994.

NATIONAL TELECOMMUNICATIONS COUNCIL

During the time when the company was being created, a regulatory body was also in the process of being created which was to be responsible for undertaking tasks of sovereign nature which were carried out by STPC previously. In the interim, and until the full establishment of this body, STPC is still continuing to perform these functions.

The law enacting a regulatory body named National Telecommunication Council was passed in November 1993 and the body is operational.

The main functions, features and responsibilities of the regulatory body can be outlined as follows:

☐ ensure that the targets of the National Strategy in the field of telecommunications are implemented as planned by the various sectors;

☐ set up rules and regulations that govern the establishment and provision of telecommunication services to the public;

☐ establish standards and norms for imported, assembled, manufactured or exported equipment from the country;

☐ ensure that a fair competition is ensured between the various entities providing telecommunications services to the public;

☐ agree on tariff structure and rates in coordination with the various concerned organizations and the company;

☐ ascertain that issues of national security and sover eignty are well preserved;

☐ ensure that major representation of the Government in international and regional organizations and conferences is maintained;

☐ licensing, provision of equipment, services and licensing of frequencies to the various private companies, governments and organizations and individuals.

The National Telecommunication Council will be headed by a Secretary General responsible to the Minister of Communications and Tourism. The Board of Directors is comprised of representatives of various Government departments concerned, the Ministry and the private sector. The Board of Directors is to be chaired by the Minister of Communications and Tourism.

This regulatory body will have a number of departments, namely:

❖ the Planning Department;
❖ the Technical Department;
❖ the Administrative and Financial Department.

The Council's personnel should be competent in the field of telecommunications and the majority of the work is expected to be implemented through committees. At the initial stages, the thrust of the work will be that of laying down rules and regulations governing the provision of telecommunication services to the public by the various entities. Monitoring and licensing will be one of its important and vital tasks.

The budget for this council is provided by the Central Government and from the fees it takes from the service providers, as well as from importation and frequency use licenses.

This regulatory body is expected to be strong enough and technically equipped to ensure that an ordely development of telecommunication services is effected throughout the country and that the rules and regulations laid down by it are implemented or are adhered to by all concerned.

WHAT IS LEFT OF SPTC?

It must be stated that it may not appear normal to have two entities providing telecommunication services within the same country when speaking of a developing country where the telecommunication market is yet to be developed. It may not also appear normal to have a public company providing service in the commercially viable areas, while the remaining Government entity is left with the rural areas whose revenues cannot support expenses of operation and maintenance and staff salaries. As has been mentioned before, in order to encourage the private sector to invest, the above course of action has been followed hoping that in the near future the rest of the country may merge gradually into the company.

What is left of SPTC will be responsible only for those the rural areas which are outside the zone of the Capital and outside the path of microwave links. In practice, most of the rest of the country will served by Sudan Domestic Satellite System (SUDOSAT) representing the only major transmission facility left to STPC. The exchanges and local cable networks are either very old or non-existent.

The required investment is huge, but nevertheless is remains a question how the Government will support financing of projects in unserved rural areas, especially when there is no clear mandate of how the Government will set aside a portion of its revenues for development of telecommunications in the rural areas.

The fact that there is no clear commitment by the Government in how it will support or promote rural telecommunication development led to some of the State Governors to think seriously in establishing their own telecommunication companies with the participation of the local private sector instead of waiting for the company to develop the major commercial centers first.

We are yet to see how privatisation as structured in the Sudan will perform and give positive results. The issue of the large staff remaining to STPC after the new company recruits its staff, which are expected to be very few, is an issue which needs to be resolved with care and wisdom. At the outset of the restructuring process, SPTC had 7,000 staff. Sudatel is employing 2,100 staff, largely taken from SPTC. The Government has to very quickly decide how it is going to support STPC in meeting its operational, maintenance and development obligations.

The new administrative structure has to be tailored to match with the new size, nature and obligations of STPC. STPC has to work on a commercial basis and has to give similar or equal incentives to that given by the company if its duties and responsibilities has to be quickly transferred to the company and/or dissolved into local state telecommunication companies.

CONCLUSION

It must be stated here that privatisation in the developing countries and especially in Sudan has been implemented with the main objective of attracting investments by the private sector in the telecommunication sector and improving performance, rather than for diversifying the carriers or services or introducing competition.

It must also be stated that privatisation is the only answer for the quick development of telecommunications in the developing countries with scare foreign currency earnings.

It is also clear that privatisation of the telecommunication sector in developing countries with all the revenues and benefits it brings to the Government, needs a strong political will and a lot of technical and administrative work in order to achieve it.

It must also be stated that the telecommunication market in most the developing countries can't be considered big enough to support more than one carrier as is the case in most of the developed countries, so that the full benefits of privatisation can not be entertained. Therefore, care has to be taken not to shift from Government monopoly into a private company monopoly without benefitting from competition, an important ingredient of privatisation.

The issue of involvement of multinational telecommunication companies from developed countries in the telecommunication development in the developing countries is still looked at with skepticism in most of the developing countries which are dominated by the security aspects of telecommunication rather than its commercial importance.

Privatisation of the telecommunication services in the Sudan must be considered as a very important positive step towards the right direction. Despite of the difficulties encountered in the process they do not rank to the level of disadvantages of privatisation rather than differences in methodologies and approaches to privatisation.

The establishment of a National Telecommunication Council is another important step towards the orderly development of telecommunications within the whole country.

Demonopolisation of purchase of the terminal equipment (telephone sets, fax machines, telex machines, PBX's, secondary and primary cables) by STPC has been one of the encouraging elements for privatisation.

Although monopolisation of international traffic by SUDATEL for over 15 years can not be considered to be in resonance with the essence of privatisation, deregulation, liberalisation, demonopolisation, yet this should not prevent us from pursuing improved telecommunication services, through diversifying networks, diversifying service providers, diversifying the types of services provided, optimizing the use of the wide band back-bone networks through VANS.

The Government of the Republic of Sudan has privatised most of the commercially viable government sectors and telecommunication has been one of them. Most of the Government Ministries are under transformation into regulatory organs or functions, so we will be seeing this kind of evolution in most of the developing countries in the few years to come.

OoOoOoO

* Based on a presentation at the conference Telecommunications and Economic Prosperity - a Strategy for Africa, organised in cooperation with the World Bank, 21 - 25 February 1994, Abidjan, Ivory Coast.

APPENDIX

TELECOMMUNICATIONS FOUNDATION OF AFRICA

OBJECTIVE:

Telecommunications sector development - creating better conditions for telecommunications development

Activities:

❖ Investment promotion
❖ Management development seminars
❖ Publishing
❖ Market research

Technical overhaul, liberalisation and sometimes privatisation of telecommunications operating companies feature very high on the agenda of many countries in Africa. This merits the conclusion that the moment has arrived for the industrialised world to take a greater interest in the African telecommunications market.

The telecommunications facilities of countries in the Sub-Saharan region of Africa are lagging far behind those in any other continent, in terms of quality, quantity and diversity. In the early 90s, no large scale schemes were in place to substantially upgrade these facilities in the region, as has been the case for most other regions of the world. At the root of this deadlock in development prospects lie insufficient absorption capacity of technology, lack of managerial skills, a relatively intransparent telecommunications market, and last but not least, favourable conditions for international firms to invest in other regions first.

The Telecommunications Foundation of Africa (TFA) considers telecommunications to be an enabling technology for business, educational, governmental and social activities. This explains the involvement of users of telecommunications in its activities.

LINKING THE INDUSTRIALISED AND THE DEVELOPING WORLD

TFA was founded in 1992, in order to assist in creating better conditions for enhancement of telecommunications facilities in Africa. TFA is an international, independent organisation and operates on a non-profit basis. The Foundation is firmly rooted in both the industrialised and the developing world.

As its main objective, TFA intends to further the development of telecommunications networks for both the private and public sectors in all countries of the Sub-Saharan region in Africa. Some of the activities of the Foundation are aimed at improving conditions for international parties to start activities or to widen the scope of their activities in the region.

INTERNATIONAL FUNDING AND MEMBERSHIP

For funding of its projects, TFA approaches international development organisations, development cooperation divisions of individual countries' governments, non-governmental organisations, as well as private companies. The Foundation also derives funds from coordinating or executing research assignments and sales of publications. TFA non-voting members represent operating companies, systems suppliers and financial organisations. Members pay annual fees.

ACTIVITIES

TFA has selected the following activities as being essential for the achievement of its main objective:

(a)   annual (bilingual) training conference on management and telecommunications, to be held in a Sub-Saharan African country;

(b)   dissemination of course material among telecommunications operating companies and relevant government divisions;

(c)   regular publication of organisational profiles of telecommunications operating companies and their respective government divisions;

(d)   organisation of briefings with potential investors from industrialised countries, on projects which are defined by telecommunications operation companies and their respective government divisions;

(e)   publication of books as well as briefs on specific subjects;

(f)   any other development cooperation activity in line with the main objective of TFA.

Any profits realised by TFA on any of its ongoing activities will be used to reinforce the financing of other activities.

BOARD AND MANAGEMENT

As of 1 January 1994 the Board of Governors of TFA includes:

(a)   Mr. Marcel C.M. Werner, chairman and secretary (Belgium)

(b)   Mr. Bethuel A. Kiplagat, former Permanent Secretary in the Ministry of Foreign Affairs and International Cooperation, member (Kenya)

(c)   Dr. Christian Schwarz-Schilling, former Minister for Posts and Telecommunications, member (Germany)

(d)   Mr. Ernst O. Weiss, vice chairman of INTUG Europe (International Telecommunications Users Group), member (France)

(e)   Mr. J. Theo Loth, treasurer (The Netherlands)

The TFA Board members participate on a personal basis. Board members are selected either on the criterium of their professional roles in the telecommunications industry and in the area of international cooperation policy, or on the criterium of their capacity to participate in work related to Foundation projects.

Deliverables of TFA are usually in English and in French. TFA was established in 1992 as a "Stichting" (foundation) under the Law of The Netherlands and registered at the Chamber of Commerce in Amsterdam (The Netherlands).

TELECOMMUNICATIONS FOUNDATION OF AFRICA

CONTACT ADDRESSES

Zoniënboslaan 21
B-3090 Overijse (Brussels)
Belgium
tel: +32 2 657 3519
fax: +32 2 657 3848
tlx: 61 344 CONTA B

IOS Press
Van Diemenstraat 94
NL-1013 CN Amsterdam
The Netherlands
tel: +31 20 638 2189
fax: +31 20 620 3419

AUTHORS' INDEX

All authors are contributing on a personal basis. Their affiliations and addresses are listed below.

Author   Page

Kordjé **BEDOUMRA**,   3
Chief Engineer Telecommunications,
Department of Infrastructure and Industry Northern Region,
African Development Bank
01 B.P. 1387, Abidjan 01, Côte d'Ivoire
Tel +225 20 45 30, fax +225 20 49 86

Cornelis **BERBEN**,   59
European Commission, DG XIII, BU 9 4/173
Ave de Beaulieu 9, B-1160 Brussels, Belgium
Tel +32 2 296 8641 / 8639, fax +32 2 296 9132

Steven W. **BLEES**,   133
University of Amsterdam, the Netherlands,
Faculty of Communication Science

Rui **FERNANDES**,   231
Chairman and Managing Director,
Telecomunicaçoes de Moçambique
PO Box 25, Rua da Sé No. 2, Maputo, Mozambique
Tel +258 1 431 921, fax +258 1 431 944 / 421 944

Dr. Paul **GODARD**,   111
Project Director, Gondwana Project
Université Catholique de Louvain
5/14 pl. Croix du Sud, 1348 Louvain-la-Neuve, Belgium
Tel +32 10 47 34 68, fax +32 (10) 47 34 71
e-mail: godard@bota.ucl.ac.be

Geoff **HAINEBACH**,   171
Joint Managing Director, Siemens Limited
P.O. Box 912469, Silverton, 0127 South Africa
Tel +27 12 836 2106, fax +27 12 836 2078

Odd **HAUGAN**,   31
Senior Project Manager, Telecommunications,
European Bank for Reconstruction and Development (EBRD)
One Exchange Square, London EC2A 2EH, United Kingdom
Tel +44 71 338 6000, fax +44 71 338 6674

Kees W. **HOZEE**,   87
Manager Carrier Relations, S.W.I.F.T.
Avenue Adèle 1, B-1310 La Hulpe, Belgium
Tel +32 2 655 3400 / 655 3111, fax +32 2 655 3226

Ernest Olumide Blyden **JOHNSTON**,   279
Deputy General Manager, Sierra Leone External Telecommunications Ltd
7 Wallace Johnson Street, P.O. Box 80, Freetown, Sierra Leone
Tel +232 2222 3424, fax +232 2222 4439

Reward **KANGAI**,   161
Director Manufacturing and Projects,
Posts & Telecommunications Corporation
P.O. Box 8061, Causeway, Harare, Zimbabwe
Tel +263 4 728 811, fax +263 4 731 980 / 730 813

Dr. Tim **KELLY**,   125
Head of Operations Analysis, Strategic Planning Unit,
International Telecommunication Union (ITU)
Place des Nations, CH-1211 Geneva 20, Switzerland
Tel +41 22 730 52 02, fax +41 22 730 58 81

Hans **KONING**,   133
University of Amsterdam, the Netherlands,
Faculty of Communication Science

Etienne **KONAN** Kouadio,   71
Deputy Director, Legislation and Regulation,
Ministry of Transports and Telecommunications
Immeuble Postel 2001, 21e étage Porte 04, Abidjan 01, Côte d'Ivoire
Tel +225 34 73 53, fax +225 34 73 68

George R. **LANGWORTH**,   147
Director of Marketing, AFRICA
FLAG Limited
c/o Nynex, Nynex Systems Company, 222 Bloomingdale Road, White Plains NY
10605 USA
Tel +1 914 4323, fax +1 914 644 4345

Richard **MAGA**,   221
Director Centre d'Etudes des Télécommunications du Cameroun (CETCAM)
Immeuble Beaulieu, Yaoundé, Cameroun
Tel +237 223 050, fax +237 231 663

Mmasekgoa **MASIRE-MWAMBA**,   259
Group Manager Commercial,
Botswana Telecommunications Corporation
Khama Cresent, PO Box 700, Gaborone, Botswana
Tel +267 358 000, fax +267 313 355

Michael **MINGES**,   11
International Telecommunication Union (ITU) -
Telecommunication Development Bureau (BDT)
Place des Nations, CH-1211, Geneva 20, Switzerland
Tel +41 22 730 55 19, fax +41 22 730 64 49
e-mail: Internet: minges@itu.arcom.ch

Pinky **MOHOLI**,   79
General Manager INTOUCH programme, Telkom
Private Bag X 74, Pretoria 0001, South Africa
Tel +27 12 311 3478, fax +27 12 323 3310

Rufus **MUIGUA**,
Assistant General Manager Telecommunications Services,
Kenya Posts & Telecommunications Corporation
P.O. Box 30301, Nairobi, Kenya.
Tel +254 2 227 401 Ext. 3531, fax +254 2 213 225

Bakary K. **NJIE**,   251
Managing Director,
Gamtel Company Limited PO *Box 387, Banjul, Gambia*
Tel +220 22 88 22, fax +220 22 66 99

Dr. Christoff **PAUW**,   187
Programme Manager Telecommunications Systems, CSIR
P.O. Box 395, Pretoria, 0001 South Africa
Tel +27 12 841 3053, fax +27 12 841 4720
e-mail: cpauw@mikomtek.csir.co.za

Reginald R. **TEESDALE**,   177
Managing Director Sestel SA, B.Sc. Chartered Engineer,
Fellow IEE, Member IEEE
F - 41400 Faverolles-aux-Cher, France

Dr. Herbert **UNGERER**, 59
European Commission, DG XIII/A/1.
Avenue de Beaulieu 9, B-1160 Brussels, Belgium
Tel +32 2 296 8623 / 8622, fax +32 2 296 8391

Kari **VAIHIA**, 161
Consultant, Telecon Ltd / Telecom Finland
Telecommunications Research Centre, P.O. Box 145, 00511 Helsinki, Finland
Tel +358 204 01, fax +358 204 023 20

Ernst Otto **WEISS**, 97
Vice Chairman International Telecommunications Users Group (INTUG)
31, rue des Bois, F - 41400 Montrichard, France
Tel +33 543 22 878, fax +33 543 272 60

Dr. Rudi **WESTERVELD**, 199
Lecturer, Telecommunications Section,
Technische Universiteit Delft
P.O. Box 5031, 2600 GA Delft, the Netherlands
Tel +31-15-782678, fax: +31-15-781774
e-mail: j.r.westerveld@et.tudelft.nl

Dr. Idris Ahmed **YOUSIF**, 285
Deputy Director General,
Sudan Telecommunications Public Corporation
P.O. Box 1130, Khartoum, Sudan
Tel +249 11 77891, fax +249 11 81096

Dr. Dimitri **YPSILANTI**, 47
Head of Division Information Technology,
Organisation for Economic Cooperation and Development (OECD)
2, rue André Pascal, 75775 Paris Cedex 16, France
Tel +33 1 452 494 42, fax +33 1 45 24 9332

ACDI	Agence de Coopération et de Développement International, Canada
ACSnet	Australian Computer Science Network, Cooperative, Australia/New Zealand
ADB	African Development Bank
AMPRNE	Amateur Packet Radio Network, research, USA
APC	Association for Progressive Computing, USA
ARPANET	Advance Research Projects Agency NETwork, USA
CGIAR	Consultative Group on International Agricultural Research
CIAT	Centro Internacional de Agricultura Tropical
CIDA	Canadian International Development Agency
CIRAD	Centre Internat. de Recherche en Agronomie pour le Développement, France
CISAFA	Comm. & Inform. systems in Sustainable Agriculture & Forestry Africa
CNR	National Council of Research, Italy
CSNET	Computer Science Network, USA
DDN	Defense Data Network, USA
ELCI	Environment Liaison Centre International, Kenya
ENSTINET	Egypt National Scientific & Technological Information Network, Egypt
FAO	Food and Agriculture Organisation
FNRS	Fond National de la Recherche Scientifique, Belgium
FRD	Foundation for Research and Development, South Africa
HEPnet	High Energy Physics Network, worldwide
IBSRAM	International Board for Soil Research and Management
ICARDA	International Centre for Agricultural Research in Dry Areas
ICIPE	International Centre of Insect Physiology and Ecology
ICRAF	International Centre for Research in Agroforestry
ICRISAT	International Crops Research Institute for Semi-Arid Tropics
IDRC	International Development Research Centre, Canada
IFDC	International Fertilizer Development Centre
IIE	Institute for International Education
IIMI	International Irrigation Management Institute
IIRSDA	Inst. International de Recherche Scientifique et Développement en Agriculture
IITA	International Institute of Tropical Agriculture
ILCA	International Livestock Centre for Africa
IPPNW	International Physicians for the Prevention of Nuclear War
IRRI	International Rice Research Institute
ISFRA	Institut Supérieur de Formation et de Recherche Appliquée, Mali
ISRA	Institut Scientifique de Recherche Agronomique, Senegal
ISRA	Institut Sénégalais de Recherche Agricole, Senegal
JANET	Joint Academic NETwork, UK
JUNET	Japan Unix NETwork, Japan
MANGO	Microcomputer Assistance for NGO, Zimbabwe
MILNET	MILitary NETwork, USA
NSFNET	National Science Foundation Network, USA
PRNET	Experimental Packet Radio Network, USA
UNDP	United Nations Development Programme
UNECA	United Nations Economic Council on Africa, Ethiopia
UNEP	United Nations Environmental Programme
UNICS	University of Nairobi, Information and Computer Service, Kenya
UNITAR	United Nations Institute for Training and Research, Switzerland
USAID	United States Agency for International Development
VITA	Volunteer in Technical Assistance, USA
WARDA	West African Rice Development Association
WHO	World Health Organisation

Telecommunications and Development in Africa

erratum

Paul Godard/Chapter 10: Africa and Science–The Availability of Computer Communications

insert on page 124:

FUTURE DIRECTIONS

The first steps in setting up international computer networks include the sensitisation of institutes to international networking and related human resources. Technologies suitable for local conditions must be emphasised: Fidonet as an entry level computer network in those places where computerisation is weak; the UUCP computer network in institutions with sufficient computing skills; and the building of national links to the Internet.

Expansion of the direct dialling public telephone network will continue to be the major development focus in Africa. However, priority should be given to X.25 public datacommunications. In most of Africa, these are either non existent or inadequate to support transnational computer communications. This puts African research at a severe disadvantage.

Meanwhile, alternative means such as satellite and radio deserve a better place in national network plans, as they can provide network access to poorly served locations. Satellite links to low earth orbit satellites (LEOS) acting as flying *store–and–forward* mailboxes can be provided economically. The earth station could cost as little as US$ 5,000. HF and VHF radios providing real–time message delivery could cost around US$ 3,000.

GONDWANA

The Gondwana project intends to inform and train African scientists in using affordable electronic tools. The Gondwana team will establish communication links with international networks for individual researchers chosen among the African Association for Biological Nitrogen Fixation (AABNF), in an expedition crossing 20 countries. AABNF was chosen in view of its presence and dynamism in nearly every African country. The Gondwana project will contribute to the development of a continent–wide network for decision makers, scientists, technicians and trainers working for sustainable development in agriculture and forestry (CISAFA).

References:
* *The Matrix, Computer Networks and Conferencing Systems Worldwide.* John S. Quaterman. 1990. Digital Press.
* *Electronic Networking in Africa. Advancing Science and Technology for Development. Proceedings of Workshop on Science and Technology, Communication Networks in Africa.* 1992. The African Academy of Sciences & The American Association for the Advancement of Science, Nairobi, 27–29 Aug 1992.
* *RINAF.* 1992. Proceedings of the Launching Meeting of the Regional Informatics Network for Africa, Dakar 27–29 Feb 1992. UNESCO & ACCT.